KT-513-880

BRING ME
SUNSHINE

Also by Charlie Connelly

Stamping Grounds: Liechtenstein's Quest for the World Cup

Attention All Shipping: A Journey Round the Shipping Forecast

In Search of Elvis: A Journey to Find the Man Beneath the Jumpsuit

And Did Those Feet: Walking Through 2000 Years
of British and Irish History

Our Man in Hibernia: Ireland, The Irish and Me

BRING ME SUNSHINE

A Windswept, Rain-Soaked,
Sun-Kissed, Snow-Capped
Guide to Our Weather

CHARLIE CONNELLY

Little, Brown

LITTLE, BROWN

First published in Great Britain in 2012 by Little, Brown

A CIP catalogue record for this book
is available from the British Library.

ISBN 978-1-4087-0324-3

Typeset in Bembo by M Rules
Printed and bound in Great Britain by
Clays Ltd, St Ives plc

Papers used by Little, Brown are from well-managed forests
and other responsible sources.

MIX
Paper from
responsible sources
FSC
www.fsc.org   FSC® C104740

Little, Brown
An imprint of
Little, Brown Book Group
100 Victoria Embankment
London EC4Y 0DY

An Hachette UK Company
www.hachette.co.uk

www.littlebrown.co.uk

For Louis

CONTENTS

AUTHOR'S NOTE

We are all experts on the weather yet none of us are experts on the weather, and *Bring Me Sunshine* attempts to bridge that gap. When I began this book my meteorological know-how extended no further than an inclination to stay indoors when it rained. *Bring Me Sunshine* has taken me on a meteorological odyssey that encompasses geniuses, obsessives, crackpots, charlatans, dreamers, storms, blizzards, gales and pea-soupers to ensure that whenever I look out of the window in the morning, tune in to the shipping forecast or dare to venture out without a coat I'll never take the weather for granted again.

I'll never take for granted the assistance of those who've helped put this story together either. I'm immensely grateful to Richard Beswick and Zoe Gullen for their help, guidance and encouragement from inception to execution, and to my agent Lizzy Kremer for not rolling her eyes and tutting at the original idea. I'm also grateful to Keith Lambkin, Rosita Boland, Dr Caroline Dodds Pennock and Bob Johnston, supremo at the Gutter Bookshop in Dublin, for their advice, assistance and wisdom along the way. Thanks also to Bernard Sumner for a much needed and perfectly timed late-night pep talk. Most of all, enormo-megathanks and all my love to Jude for yet more of her unshakeable, smiling love and faith, without which I just couldn't get up in the morning, let alone check what the weather's doing.

If *Bring Me Sunshine* somehow isn't enough Connelly for you, you can find me at my website www.charlieconnelly.com, on Facebook at www.facebook.com/charlieconnellyauthor and on Twitter at @charlieconnelly.

1

A SLOW DANCE OF SUN AND RAIN: THE WEATHER AND US

I see quite a lot of our postman. I work from home and receive a fair number of parcels, meaning he and I have become cheerily familiar and have a well-worn routine whenever he calls. I go and open the door, he smiles and says, 'Morning', I smile and say, 'Hey, how's it going?', then he rummages through his armful of envelopes for whatever it is that won't fit through the letterbox.

It's in this *après*-greeting stage of the situation that awkwardness could set in, for this is a potential minefield of modern etiquette. We're undertaking what is essentially a business transaction: he's providing me with a postal delivery service and, familiar though we are, it's not like we're poking each other on Facebook and going to the match together at weekends or anything. Having said that, he knows my name, my address and that I subscribe to more magazines than I have the time to sit down and read. He knows I generally swan around in the mornings wearing a pair of Charlton Athletic shorts and an old Beach Boys

T-shirt because that's what I'm usually wearing when I answer the door. He knows the name of my girlfriend and where we have our bank accounts, and he could have an accurate stab at when our birthdays are – all information he can glean from the daily ritual of putting stuff into our letterbox. I, on the other hand, don't know anything about him other than I think his name might be Neil and that he goes to Turkey for his holidays.

So when he knocks with a parcel and we've exchanged hellos it's a textbook set-up for a bit of mild social tension. What do we say to each other that keeps the relationship cheerful and informal but not intrusive? How do we fill these seconds that have us teetering on the precipice of embarrassment? Luckily I have a foolproof way of defusing any awkwardness and it's one that I'm quite certain you're familiar with. I rub my hands together, look at the sky, squint slightly and say one of the following:

'Hope we've seen the last of that rain.'

'Not as cold today, is it?'

'Oof, that wind'll go right through you.'

'It's supposed to stay like this for the weekend, apparently.'

Neil the postman flat-bats a suitable response back to me, hands me my parcel, and we bid each other cheery farewells until the next time.

Thank goodness for the weather.

That we talk a great deal about the weather is one of those copper-bottomed clichés underpinning our national identities. It props up our sense of self-esteem that when we fall into conversation with a stranger at a bus stop we have an inoffensive meteorological discussion instead of, say, asking if they've ever had a yeast infection, or trying to tickle them.

These conversations are never enlightening, either. They'll rarely get to grips with curiosities like the time in June 1975 when Derbyshire played Lancashire at Buxton and there was a blizzard, the only instance of 'snow stopped play' in the history of first-class cricket, or the fact that the French Revolution was triggered by a 1788 hailstorm in northern France that destroyed an already meagre harvest, led to food shortages and civil unrest and, ultimately, the storming of the Bastille. Usually, in these conversations we're complaining about the weather. Most of the time we're suspicious of it; certainly we're very distrustful of it. We're usually suspicious in our responses to it too: one sunny day, and we're counting the minutes until the next hosepipe ban; a flurry of snow, and we're predicting instant chaos on the railways due to the 'wrong kind of snow'. On very rare occasions we are actually pleased about it but always with the shrugging caveat that it can't possibly last. It's our way of bonding with a stranger, of putting ourselves on the same side. Neither of us wants to sit or stand there in awkward silence but neither of us wants an in-depth conversation about matters of great import either. We want to find ourselves on common ground, agreeing about stuff of little consequence and then going our separate ways with the other thinking, Well, he was a nice bloke.

In a way it's like a particularly lame version of the old Blitz spirit: we're united in our scanning of the skies, only these days we're looking for flecks of grey in the clouds instead of the bombs our grandparents were expecting to plop down their chimneys at any moment. Yet despite our antipathy to our climate, the fact is that our weather is, in the great scheme of things, not that bad. Honestly, it really isn't. In Britain and

Ireland we're among the relatively few people in the world whose weather is not susceptible to great extremes, is reasonably straightforward to predict and allows us a pretty comfortable and worry-free existence. When our weather is notable it's generally because it's a bit *more* of something than usual, rather than the kind of blockbusting meteorological event that has Hollywood producers picking up the phone and dialling Will Smith's agent. We don't suffer monsoons, life-threatening droughts, tornadoes that leave just a few kitchen-floor tiles as a clue to where our house used to be, blizzards that last for months or sandstorms that can blot out the sun for days. Our winds are not so severe that we risk going outside and being hit from above by a cow; our rain not so heavy as to have a pod of pilot whales looking at us through our living-room windows.

On the face of it our weather seems almost entirely inoffensive. If weather were literature ours would be written by P. G. Wodehouse; if it were rock 'n' roll we'd be represented by Bananarama. Our notable weather events are more about breaking minor records than lists of casualties. Most of us remember where we were on 10 August 2003, for example, when Britain's highest ever temperature was recorded: 38.5° Celsius, or 101.3° Fahrenheit. I was in the Western Isles of Scotland. Stuck in a thick sea fog. Wearing an extra fleece.

We're all about records: barely a month seems to go by these days without it being the somethingest since records began. In October 2011, for example, Gravesend recorded Britain's highest ever October temperature, a sweltering 29.9°C. And how very British it was to stay just under the thirty-degree line: ticking over to thirty would, of course, have just been vulgar.

Granted, we do have the odd insurance-premium-hiking weather event like the startling flash–flooding in Boscastle in August 2004, when warm, damp air coming in off the Atlantic being forced upwards and cooled quickly by the shape of the landscape led to sudden torrential rain that flowed down from high ground above the town, caused the River Valency to burst its banks and wash away a bridge and several cars, and caused flooding to a depth of nearly three metres that damaged a hundred homes and businesses. We still think of Boscastle as one of our major recent weather events, but even then the only human casualty was a broken thumb.

It's the rarity of these events that makes them notable. Our weather is generally steady, reliable, constant and unspectacular: exactly why it's perfect conversational fodder for the envelope-shuffling moments that occur between me and the postman.

These weather-based conversations are perfect because they never provoke confrontation. Think of all the times you've said to someone, 'Isn't it a lovely day?' How often have they responded, 'What are you, some kind of nut? It's a *rotten* day'? These exchanges are nothing more than silence-fillers designed to pass time rather than engage. We use the weather because it's possibly the only piece of common ground we can be sure we all have, other than going to the toilet, and we're certainly not going to quiz strangers about that. The weather is a common bond; we all spend our days walking around under it and we all pretty much agree on whether it's good or bad. Even if we venture to make a prediction – 'I think it might rain later' – there are no consequences if we're wrong. No one has ever been sued over a bit of meteorological small talk.

The superficial nature of these conversations proves that,

despite how much we might talk about the weather, we don't know all that much about it. Some people say we're obsessed with it: this isn't necessarily true. Ask someone the lowest recorded temperature in Britain and Ireland and I bet they wouldn't know it's −27.2°C, recorded twice in Braemar, Aberdeenshire, in 1895 and 1982, and once − in 1995 − at Altnaharra in the Scottish Highlands. Ask them the highest recorded wind speed and they'll struggle to come up with the 142mph gust recorded at Fraserburgh in 1995. We might talk about the weather often, but it's only ever in terms of the massively obvious: it's a nice day, it's clouded over, it's getting a bit windy, it's turned colder. This is not the language of the obsessed. If we're obsessed with anything it's avoiding awkward moments, not the weather. We have a fairly limited vocabulary to describe current conditions because with a few important exceptions − farmers and seafarers, for example − we don't need to know much about the weather. Beyond the odd minor inconvenience we don't depend on it, as most of our existence is spent indoors keeping out of its way.

Yet we *should* be obsessed with the weather because it's vital to understanding who we are and where and how we live. The weather has been ingrained in our collective consciousness from the earliest times because we developed as a farming and maritime nation. Dr Johnson is usually quoted at this point ('When two Englishmen meet their talk is of the weather'), but we can go even further back:

> *Whan that Aprille with hise shoures soote,*
> *The droghte of March hath perced to the roote,*
> *And bathed every veyne in swich licour*

Of which vertu engendred is the flour;
Whan Zephirus eek with his swete breeth
Inspired hath in every holt and heeth . . .

Geoffrey Chaucer, writing in the late fourteenth century, began the General Prologue of *The Canterbury Tales* by describing how the arrival of April showers saw off a dry March as Zephyrus, the west wind, breathed life into every grove and field. The first lines of the first great work of literature in our culture talk about the weather. It's not just an opening picked at random: Chaucer's use of the weather places those gathered in the Tabard Inn at a particular time of year, but also it gives him an opening with which people from all levels of society could identify – a common experience everyone understands. It is also an arresting opening packed with beautiful imagery: the sweet breath of the spring wind and the *shoures soote* – fragrant, freshening rainfall.

Chaucer's contemporary William Langland begins *Piers Plowman* with a mention of the weather, and even the best part of two centuries earlier Gerald of Wales had not got far into his *Topographia Hibernica* before mentioning how 'this country more than any other suffers from storms of wind and rain'. *Sir Gawain and the Green Knight*, also from the fourteenth century, passes the year between the two encounters of the protagonists with an evocative description of the changing weather and seasons.

So, talking about the weather is not the product of modern etiquette: we've always done it. Also, despite Oscar Wilde regarding such conversation as 'the last refuge of the unimaginative', for as long as we've told stories or kept histories the weather has inspired and influenced writers, poets and artists. It has affected the course of history too: the Normans' voyage across the English Channel being delayed by northerly winds for most of the summer of 1066; the storms that aided the scattering of the Spanish Armada; the unseasonable warm dampness that incubated the blight which caused the Great Famine in Ireland; and the millpond conditions in the Channel on D-Day.

When Neil the postman and I exchange pleasantries over the Jiffy bags we are continuing a tradition that goes back centuries and can be found in our greatest works of literature and art. Yet for most of us the weather is still, at best, a harmless topic of conversation.

Our weather deserves more than this – after all, we're either under it, in it or sheltering from it every day of our lives. It affects the way we live; it keeps us alive. When it breaks out of its regular rhythms and routines it stays in our consciousness: when we think about 1976 we might think of the coming of punk or Southampton winning the FA Cup, but equally we might think of that scorching summer, the hottest for 350 years and with the most severe drought in a millennium, when for twenty-six consecutive days in June and July the temperature reached 80°F. When people in the south of England recall 1987 they might think of the great October storm that tore up the place; similarly in Scotland and Northern Ireland 1990 might conjure memories of the Burns Day gale. In years to come the harsh winter of 2009–10 will doubtless feature in the same way for its school closures, widespread power outages and transport chaos.

But, apart from these major events and the odd localised weather disruption, we go about our business largely ignorant of the weather, its causes and its effects on our lives and our world.

If it hadn't been for a sleepless night leading to the realisation that I had fewer excuses than most for not acknowledging the importance of the weather, I might have just continued my regular happy discourse with Neil and left the weather to its own devices. Instead I found myself embarking on a mission of extraordinary discovery, delving into the elements to learn about how and why the weather happens and how we've tried to control, harness and predict it. That I embarked on this story at all is down to that iron horse of meteorology, the shipping forecast.

It was some time in the middle of the night, that no man's land where it's neither late nor early, and I'd been woken by a storm hissing through the trees and roaring around the building. I lay there for a while, turning one way and then the other, mind whirring, doing that thing where the dead of night transforms the most trivial daytime concerns into stomach-churning horror: the approaching deadline, the unanswered e-mail, the mildewing washing still in the machine; the sort of mundane 'to do' list that suddenly seems so portentous that you wouldn't be surprised if a nearby bush suddenly burst into flames and outlined the potential consequences in a booming voice.

I gave up on sleep, heaved myself out of bed, padded into my office and switched on the desk lamp, intending to read myself tired again. The wind was louder in there as the window faces the sea and was hence copping the full onslaught of the night's gusts. The glass that separated me from the elements was the first obstacle the wind had encountered in its long run-up from whatever depression in the Irish Sea had set it spinning before launching itself at the land. On my desk, just at the edge of the small pool of light cast by my lamp, I noticed the mug that holds my pencils. In daylight hours I barely notice it, but now that it lurked at the edge of my vision, half hidden in the shadows, it caught my eye. On one side of the mug there's a map of the areas of the shipping forecast and on the other, the one that faced me, the Beaufort wind scale. I read down the thirteen demarcations as they increased in severity and wondered what Force might be sweeping in and depositing spots of brine on my window. If, as it seemed, this was seawater whipped from the white-tipped waves a couple of hundred yards away then, according to the descriptions set out by Francis Beaufort, this was maybe a force 7: a near-gale, where the 'sea heaps up; white foam from breaking waves begins to blow in streaks'. It might even be a fully fledged gale, force 8, with 'moderately high waves of greater length; edges of crests break into spindrift; foam blown into well-marked streaks'.

I couldn't see the waves in the dark, so thought I'd see what the shipping forecast had made of it. With the wind crashing around outside and the sea hidden away in the darkness I heard the familiar strains of 'Sailing By', followed by the solemn, rhythmic intonation of the weather forecast for the seas around Britain and Ireland. 'There are warnings of gales in Viking, North Utsire, Forties, Cromarty, Forth, Tyne, Dogger, Irish Sea, Shannon, Rockall, Malin, Hebrides, Bailey, Fair Isle, Faeroes and South-East Iceland,' said the calm, faintly stern female voice easing quietly from the speakers. Then began the familiar ritual of the clockwise progression around the map, starting on the coast of northern Scandinavia, criss-crossing the North Sea then racing westward along the English Channel, turning south along the coasts of France, Spain and Portugal, doubling back and bearing down on the southern tip of Ireland before veering off to the north-east and arriving at my window.

'Irish Sea, Shannon, westerly or south-westerly five to seven, occasionally gale eight, occasional rain or showers, moderate or good, occasionally poor in Shannon.'

It turned out my guess, based on nothing more than a complaining ventilation panel and some salty specks on my window, had actually been pretty good. I closed my eyes and listened as the forecast made its way around the rest of the map, leaping over Ireland into the north Atlantic then skirting the western and northern fringes of Scotland before heading off over the sea towards Iceland, all to a soundtrack of the gale outside.

There was something very familiar about this scenario, a memory from a long time ago and a long way from here. I'd sat at a desk just like this, late on a rainy night with a pool of light from a lamp the only illumination, listening to the shipping forecast. Back then I was in south-east London, looking out of the window at a rain-soaked, orange-sodium-lit street and being inspired to undertake a year-long journey around the sea areas that make up the familiar yet enigmatic maritime roster. As a result, the forecast's list of names would always evoke images of real places and real people, real stories from heroism to farce, real friends I'd made who are still friends, some pretty rotten weather and some pretty good weather. Even the picture of my seafaring great-grandfather that I'd carried with me and the chunk of volcanic rock bearing the Icelandic flag that I'd bought to mark the final leg of my journey were there on the window sill.

The forecast moved on to the reports from coastal stations and it dawned on me that for all the people I'd met and the experiences I'd had on my way around the shipping forecast, for all that I knew the logic behind the mysterious terms and poetic phrasing, I knew next to nothing of the very reason it exists; what those familiar terms were predicting. I'd found out how it was written, how it was broadcast and how it came about in the first place. I'd met people who lived by it and in it but I was still entirely ignorant of its primary source: the thing that had roused me from sleep that night, the thing that frequently defined the kind of mood I'd be in, the thing that I always talked about with the postman.

As the final reports from coastal stations – Valentia, Ronalds-way, Malin Head, Machrihanish – rounded off the forecast, I realised that in missing out the weather from my homage to the shipping forecast it was as if I'd reported on a football match without revealing the score.

A few days later I was staying with some friends in a farm-house on the coast of South Wales. The room in which I slept was usually occupied by a small boy and had a shelf full of old Ladybird books. One morning, having woken up before every-one else because I was on holiday and excited, I killed a bit of time by looking along the row of Ladybirds, many of them familiar from my own childhood. One was devoted to the weather, its cover dominated by a windmill beneath a sky full of clouds, so I pulled it from the shelf and started to leaf through it. It was a classic of its type, brimming with wonderful colour illustrations of boys wearing shorts and pullovers looking in wonder at rain gauges and weathervanes while a man in a tank top looked on approvingly, a pipe clenched between his teeth and balled fists placed assertively against his hips. But when I read the accompanying text – written for kids of about eight, remember – I realised that I didn't understand any of it. Barely a word. I'd read the same page over and over again, seeing familiar words like pressure, barometer and millibar, but I was clearly miles behind the shiny-faced youngsters in the pictures who, even though trapped in their ageless wonder in an image half a century old or more, still knew far more about the weather than I did.

I looked out of the window across the marsh to where the nearest town nestled shyly between promontories and watched as a grey smudge, a smeary pennant of drizzle, glided soundlessly and slowly across the mouth of the harbour and fuzzed the outlines of the buildings. Another followed close behind before a watery sunlight broke through the clouds and sent a patch of pale yellow moving across the marsh grass towards the town, the one that had inspired Dylan Thomas's Llareggub in *Under Milk Wood*, which sparkled sharply and happily in the warmth and the freshly rinsed air and doubtless smelled of seaweed and breakfast.

I slid the book back into place on the shelf, looked out at the remnants of the meteorological dance I'd just witnessed over Laugharne and realised something that had never really struck me before: the weather is beautiful. From rainbow to rainstorm, crisp frost to big wet flakes of snow billowing through the sky, gusts of wind ribbing in waves across fields, spotless white ranges of cumulonimbus clouds towering into the sky, and sublime lightning storms that turn day to night and make the sky boil and rage, all of it has a deep and different beauty. Even that brief shower on the other side of the marsh travelled with a wonderful silent grace across the sea, motion-less yet always moving, brushing over the town like passing silk.

Sometimes it's not much fun being beneath the weather when it's emptying large quantities of water over you or whipping your hat from your head and sending it bowling across the road like a Victorian child's hoop. I've had more than enough drenchings to last me a lifetime, be they at football grounds or halfway up mountains while consulting a soggy map, but it was only seeing those faint smudges against the sky and the stately progress of the sunshine across the fields from the upstairs window of an ancient Welsh farmhouse – a view that had remained unchanged geographically since the house was built but that was reworked by the weather, often by the minute – that opened my eyes to the real beauty of the weather. And I wanted to know more.

My surprisingly accurate perception of the Beaufort scale with the shipping forecast, the early-morning slow dance of sun and rain over a literary Welsh town as it was waking up and my embarrassing besting by an ancient kids' book all combined that morning to give me a new purpose. I would delve into the beauty and mystery of the weather, learn about each of its aspects and immerse myself in its history. I'm often asked why I think the shipping forecast has such resonance with people and I've always responded that it is because of the combination of its having been broadcast on the BBC for the past eighty years or more and the fact that we're islands, for whom the sea is the defining border.

Our relationship with the weather has a history full of incred-ible stories, amazing discoveries and extraordinary people doing marvellous things. There are heroes, villains, nitwits and char-latans peppering the story, many of whom I'd learn about on my quest to understand the beautiful mystery of the weather. But as well as an attempt to put right my inexcusable ignorance, I hoped that this quest would lead me to discover much more than what an occluded front might be and how we arrived at the barometer. I didn't know it at the time but I was about to embark on a weather journey that would include the creation of the planet itself, winds that make us mad, an epiphany as to the true wonder of the umbrella, some tremendously entertaining fraudsters, a gaggle of well-meaning philosophers, people who went to war with the sky, some of history's most spectacular storms and some startlingly brilliant Victorians. Oh, and a man who kept being struck by lightning. In addition, I hoped I might understand more about ourselves: how we hardy folk on these windswept, rain-soaked islands on the fringe of Europe and at the mercy of the Atlantic came to be who we are today. To do that there was only one place to start – the rain.

2

OCCASIONAL RAIN OR SHOWERS, GOOD

In our collective psyche the weather is almost defined by pre-cipitation. When we think of the weather we usually think of the rain: in our minds we are stuck here on these lumps of wet rock at the mercy of the rain, ranging from full-blooded Atlantic storms to a cold misty drizzle from a featureless grey sky. It's almost as if rain is our default weather setting; its inevitability in our minds is like a national ennui. Our distrust of the weather stems from the expectation that the heavens will open at any moment.

Yet we're far from exceptional in terms of rainfall and in global terms come in well below the average. According to the United Nations, in the league table of average annual rainfall Britain comes in well down the field at number 48, Ireland at number 50. In European terms the Netherlands, Norway, Iceland, Luxembourg, Germany and Italy experience more rainfall than Britain, while Portugal is sandwiched between Britain and Ireland in the global rankings (the top three, inci-dentally, in ascending order are Sierra Leone, the Solomon

Islands and Equatorial Guinea, the last experiencing five times as much rain as we do each year). There is not one British city among the top ten rainiest in Europe either. Dublin comes in at number 7. The wettest city in Europe by some distance is Zurich with 42.3 inches each year, a full four inches ahead of second-placed Milan. Believe it or not, London is the sixth *driest* city in Europe with 23.3 inches of rain each year (Athens is the driest, with fourteen). Even within Britain the myths don't stand up: ask most people what the wettest city might be and they will probably say Manchester. Yet if you tot up the figures Manchester only just makes it into the top ten: in actual fact the rainiest city in Britain is Swansea, with an average of 1360 millimetres, or 53.5 inches, a year, which seems to prove the local saying that 'If you can see Mumbles Head it's going to rain. If you can't see Mumbles Head it's already raining.'

Yet despite the apparent lack of a reason for our rain-weary demeanour we still seem to regard rain as our standard meteorological condition. Our very existence is a constant jousting match with St Swithin who, we feel, is always trying to outwit us, ruining our picnics, barbecues, music festivals, cricket matches and patio furniture. He's a crafty old goat, the patron saint of weather.

There's something almost pagan about the whole St Swithin story. Staunchly Christian though he was, he's the closest thing we have to a weather god, as tradition dictates his feast day is supposed to provide us with a portent of the weather to come for the rest of the summer:

St Swithin's Day, if thou dost rain.
For forty days it will remain.

St Swithin's Day, if thou be fair,
For forty days 'twill rain nae mare.

For someone who was otherwise a pretty run-of-the-mill saint (led a pious life, performed one recorded miracle that was, to be honest, a bit lame, and died a peaceful death that didn't involve being tied to a stick and set alight), in purely religious terms St Swithin's fame seems to far outstrip his saintly significance: his sanctified career has certainly eclipsed the achievements of his life. Wherever it is that saints gather for a sweet sherry Swithin must attract a few scornful glances. You can almost imagine St Anthony the Abbot tilting his head in Swithin's direction and grumbling about how of all the things in the patronage lucky dip he got pigs and gravediggers, and who's ever needed to know the name of the patron saint of pigs and gravediggers? St Bernardino of Siena would have rolled his eyes and assented, coughing and clearing his throat and croaking that he himself is the patron saint of hoarseness and, as a result, very few people have heard of him. Yet there was Swithin, one of the best-known saints in the book despite never having been persecuted, not having met a grisly end and only having performed one miracle, which was mid-billing music-hall conjuring fare at best.

Nobody knows exactly when Swithin was born, but it was probably not long after the turn of the ninth century during the reign of King Egbert of Wessex. He was ordained as a priest by Helmstan, the Bishop of Winchester, most likely during the late 830s, and was appointed tutor to Æthelwulf, Egbert's son, becoming a close confidant of the King at the same time. Swithin succeeded Helmstan in 852 and commenced a decade-

long period of almost nauseating piety. Unlike many of his contemporary bishops Swithin would make every journey on foot
no matter what the distance, and when he gave a banquet he
wouldn't invite political or religious dignitaries, only the poor.
During his tenure Swithin instigated and oversaw a vigorous
period of building new churches and repairing the ones that had
been ransacked and damaged during the Viking raids. So far, so
saintly.

Swithin's only recorded miracle occurred when he was
making one of his long journeys on Shanks's pony. He came
across some workmen putting the finishing touches to a new
bridge and as Swithin approached he saw a woman carrying a
basket start to cross the bridge from the other direction. For
some reason the workmen took it upon themselves to relieve
her of her load of eggs and smash every single one on the
ground.

Having witnessed this startling display of caddishness Swithin
reached the scene, walked into the midst of the chortling workmen and bent down towards the mess of broken shells and
quivering yolks. The woman and the builders then looked on
wide-eyed as every egg Swithin touched was restored to its
former glory and returned whole to the woman. Now, while
this must have been an extraordinary thing to witness, as miracles go it's pretty low-wattage.

There was to be no repeat of the egg trick or anything else
remotely miraculous before Swithin died peacefully in Winchester in July 862. And there his parochial reputation as a
pious Tommy Cooper of the Middle Ages might have
remained were it not for certain posthumous events involving
his mortal remains. He was buried according to his wishes,

which were characteristically puritanical: Swithin wanted to lie outside the sanctified area of the cathedral grounds in 'a vile and unworthy place' where his grave 'might be subject to the footfall of passers-by and to the rain pouring forth from the heavens'. His wishes were granted, a suitably grim spot was found not far from the cathedral entrance, in he went and there he would remain, being soaked and trampled on in a manner that only Swithin could have found agreeable, for the next century.

However, in 971 a new basilica was built at Winchester and, as part of this, the bishop Æthelwold took it upon himself to relocate Swithin's grave to a more worthy location inside the new building and to make him its patron saint. On 15 July of that year, just as the monks were exhuming his coffin and carrying his remains to the cathedral, the skies darkened and the heavens opened. It absolutely hammered down. Not only that, but the rain didn't stop for the next forty days, something that Æthelwold and the monks interpreted as a sign that the old saint was crying tears of sorrow from the heavens. Despite nodding solemnly in agreement that this was the most likely explanation for the deluge, it didn't seem to occur to the drenched clerics that returning Swithin to his original grave might be the decent thing to do. Instead, the dead saint was interred inside the sanctity and shelter of the cathedral and the watery nature of his exhumation and its aftermath gave rise to the legend of St Swithin's Day.

Swithin's problems weren't over yet. His tomb became one of the most popular shrines in the country and his remains premier-league relics, to the extent that at some point during the Middle Ages his head was lopped off and taken to Canterbury

Cathedral while Peterborough Abbey later acquired an arm, a level of attention that you can't help feeling Swithin would have utterly detested. Either way, the consistent downpour that heralded his exhumation and reburial was enough for Swithin to be appointed patron saint of the weather.

But is there any truth in the adage that carries his name? If it's fair on 15 July will it really *rain nae mare* for the next forty days? Likewise, if St Swithin's Day is wet do we suffer forty days of wet weather? Each year on or after 15 July there's usually at least one correspondent on a newspaper's letter page apparently debunking the myth of St Swithin's Day, but while there may not be constant rain for a month and a half or so, there would seem to be a grain of truth in the legend.

H. H. Lamb was one of the great climatologists of the twentieth century and he pointed out that British summers have usually settled into their prevailing character by mid-July: generally, we know whether we can define it as a good summer or a rotten one by that point. Lamb pointed out that in June and early July there is a general shaking-up of the atmosphere, affecting barometric pressure and the way the air circulates over the northern hemisphere. This is followed by a period when things tend to stay settled in that pattern until around the end of August, when the next significant change in the weather takes place. This brackets a period of around six weeks from mid-July: St Swithin's forty days. It may not literally rain for forty days, but if the weather is grim around mid-July it is likely to stay largely that way for the period defined by the rhyme. In this case, the science broadly backs up the myth.

Before we leave St Swithin to the muttered envy of his sanctified colleagues over the cashew nuts, it's worth pointing out a significant literary connection that also marks him out from the rest. Swithin hasn't been lauded in song, prose and poem as much as you might expect: there's a Thomas Hardy poem, 'We Sat at the Window', in which a couple stare wearily out of the window at the rain on 15 July contemplating their dead relationship, and Billy Bragg recorded a song called 'St Swithin's Day', but the strongest connection is with one of our most famous novelists.

Not many people know that the last thing Jane Austen composed was a poem on and about St Swithin's Day. Not only that, she wrote it in Winchester, in a house a hop and a skip from Swithin's last resting place. Austen had been very ill from the beginning of 1817 and pretty much confined to her bed since April. The following month she was brought from Chawton to Winchester in order to lodge near her doctor but there would be no significant improvement in her health. Although illness had caused her to abandon the novel she'd been writing under the working title 'The Brothers' (published more than a century later as *Sanditon*), she lay in her bed and dictated to her sister Cassandra a poem called 'Venta' – from Venta Belgarum, the name of the original Roman settlement at Winchester – on the occasion of the annual races held at the town's racecourse on St Swithin's Day. The noise of the drunken revelry associated with the race meeting must have been constant outside her window and hardly likely to aid any recovery, so the jaunty tone of the poem comes as a surprise in the circumstances: her brother Henry later described it as 'replete with fancy and vigour'. It chides the racegoers for forgetting the saint's day, to the extent that Swithin himself rises up to admonish them, declaring:

These races & revels & dissolute measures
With which you're debasing a neighbouring plain,
Let them stand – you shall meet with your curse in your
    pleasures,
Set off for your course, I'll pursue with my rain.

Little more than two days later Austen died in her bed and was buried in the cathedral, not far from Swithin himself. The links don't end there either: Austen's parents had married in the old St Swithin's church in Bath, while her father was buried in the new St Swithin's there. In a fitting coda to the Austen–Swithin connection, the author turned out to be the last person ever to be buried inside Winchester Cathedral: appropriately a rising water table prevented any further interments.

Rain featured significantly in the lives of her characters too. In *Pride and Prejudice* Jane Bennet rides off to the Bingleys' house at Netherfield, is caught in a downpour en route, falls ill as a result and is forced to stay with the Bingleys until she recovers. This mortifies Elizabeth Bennet but delights her mother as it means Jane may catch the eye of the eligible Mr Bingley, or one of the officers also staying there. Early in *Sense and Sensibility* Marianne Dashwood is caught in a shower of rain, falls and sprains her ankle and is rescued by the dashing Willoughby, setting in train a chain of events that will break her heart. Later in the book, while mooning over Willoughby Marianne is caught in a fearsome storm, from which she catches a terrible chill and lingers on the point of death. She recovers and realises how foolishly sentimental she has been: the rain triggering both her falling for Willoughby and her eventual epiphany.

The idea of rain as a great leveller is a fairly common theme in literature. The sight of the half-crazed Lear drenched on the heath amid political as well as meteorological chaos, raging at the elements –

Blow, winds, and crack your cheeks! Rage! Blow!
You cataracts and hurricanoes, spout
Till you have drench'd our steeples, drown'd the cocks!

– is one of the most memorable in Shakespeare. Jonathan Swift's 'A Description of a City Shower' tells how

. . . various kinds, by various fortunes led,
Commence acquaintance underneath a shed.
Triumphant Tories and desponding Whigs,
Forget their feuds, and join to save their wigs.

Rain is no respecter of class, status, money or fame because it predates all of those things and will outlive them too. Maybe that's why rain figures so strongly in our culture: it is democratic, and no respecter of reputation. It reminds us not to get above ourselves and its sagacity has a long legacy: the rain has a wisdom that comes with age.

Rain was here long before we were: there's been rain for as long as there's been water in the atmosphere and a process of evaporation. Geologists broadly agree that our planet is around four and a half billion years old and that it had a troubled childhood: for its first billion years or so there were some pretty hot gases swirling around the place and the young planet was bombarded with lumps of rock and all sorts of debris flying about the fledgling solar system from the exploding star that began it all. There were constant volcanic eruptions all over the young planet, the acne of geology, but eventually after roughly a billion years – I think it was on a Thursday – the earth began to cool and vast amounts of steam and water vapour formed in the atmosphere. This vapour gathered around specks of ash and dust to become droplets of water. And then it began to rain. It rained solidly for hundreds of millions of years, forming the oceans and hammering down on steaming rocks and mantle, on volcanic ash and molten lava. It washed minerals and proteins into the sea, and these materials would eventually combine to spawn life on the young planet.

Everything, every living thing, including us, was born of rain.

About 1.6 billion years ago it rained on the area that now forms the Vindhyan mountain range in Madhya Pradesh in the north-east of India. We know this because Chirananda De, a geologist with the Geological Survey of India, found some of the raindrops in 2001. Well, not the actual raindrops, but the impressions they left in sandstone sediment exposed to the air in the lower regions of the range: circular indentations three to five millimetres across, like when rain falls on wet sand. Some are circular, others more elliptical, meaning the rain that particular day was coming straight down or at a slight angle: a shower occasionally nudged by a breeze.

If we travel back another billion and a bit years we find that it was raining in what is now South Africa. It rained onto a layer of cooling volcanic ash but it didn't rain for long: just enough time for the tiny craters to form and harden and stay hardened for the next 2.7 billion years. A heavier downpour, and any definition would have been battered out of sight; torrential rain and the ash itself would have been washed away. It was just a shower that left little doughy pustules in the ancient, yellowy-brown ash as it dried as hard as rock. Some of the oldest fossils ever discovered, much older than the curled-up ammonites and fishbone trilobites that turn up amid cliff falls and on beaches, are showers of rain frozen in time; one nondescript moment in a single place, a place that would have moved and shifted and drifted as the continents separated and formed over millions of years, just one shower of rain like the countless showers that fall every day across the globe today. A few dozen raindrops that fell long before life began.

For those hundreds of millions of years of rain there was nobody around to get wet and complain about it. Eventually we heaved ourselves out of the primordial fluid and for the first time noticed it was raining. Well, strictly speaking we weren't really clever enough to notice anything much at all back then, but those first escapees who emerged from the water and began to breathe commenced our relationship with the weather. Now that we are here and have evolved like billy-o, to the point where we can clothe ourselves, play the trombone and use cling film, it's interesting to reflect upon how much of our human development has involved working out how to shelter from the rain. Nearly four and a half million years ago, for example, 'Ardi', a hominid of the *Ardipthecus ramidus* species, lived in Aramis, a settlement in the Middle Awash region of what is now Ethiopia. She was about four feet tall, weighed roughly seven and a half stone, was covered in hair, was one of our ancestors and, well, wasn't very bright. She had a very small brain, cer-tainly compared to ours, and when you think that even with

brains our size we still create perfume adverts and vote on reality TV shows you can probably appreciate how dim Ardi
actually was. She walked crouched rather than fully upright, and
also climbed trees. She probably slept up there in order to avoid
predators – and also to shelter from the rain, having been sen-
sitised enough during her lifetime to notice that the falling
water made her cold and uncomfortable. Taking to the trees
would get her feet out of the puddles and offer shelter among
the leaves. She was found millions of years later in a layer of
mud, her bones shattered into many tiny, breathtakingly fragile
fragments, where she'd been trampled into the wet ground by
the thundering feet of larger creatures on their way to feed or
drink, and there she lay in the detritus of rain for nearly four and
a half million years.

Some of our later ancestors took to caves, looking out at the
rain through the flickering flames of fire. Eventually we constructed our own rude shelters, huts and the buildings in which
we're able to gather, live, work and socialise without getting
wet. Our powered forms of transport are enclosed and we wear
waterproof clothing and use umbrellas, meaning that it's possible for not a single drop of rain to land on us throughout an
entire day even if it's hammering down the whole time.

The rain helps us live and survive by watering crops and feed-
ing animals. The water that makes up most of our bodies would
once have fallen as rain. If it suddenly stopped raining all over
the world it wouldn't be long before our species died out
entirely. Our level of rainfall in Britain and Ireland alone is vital
to our survival: the surface temperature of the sea around us is
relatively warm and doesn't vary a great deal during the year (off
Cornwall, for example, the difference in sea temperature in

summer and winter is barely six degrees). Our islands themselves are cooler than the sea in winter and warmer in the summer because land has a lower thermal capacity than the sea, meaning the sea is slower to register changes of temperature. The temperature of the air above the sea is gradually pulled towards that of the surface of the water, and as a result we experience winters that are relatively mild, and cool winds during the summer. The air also picks up moisture from the sea to help give us our rain, rain which falls at a level that can comfortably sustain life here. If the seas around us were significantly colder there'd be less moisture in the air, less rain and not enough water for us to survive.

Yet considering this aspect of the atmosphere keeps us alive and has prompted most of our great steps forward in species development, what do we do about the rain? We moan about it. Constantly. Sometimes with good reason when it floods our homes, extinguishes our barbecues, puts a premature end to a test match or, worst of all, prompts Cliff Richard to sing a cappella to a crowd of tennis fans. Rain is good for us and in the great scheme of things we don't actually get that much of it here, yet still we complain about the rain, even though most of us experience less annual rainfall than our counterparts in Venice, Rome and even Naples. Granted they get more sunshine than we do, but in terms of rain, well, we win.

If we are anywhere close to being obsessed with anything to do with the weather, then this aspect of it is the only candidate. When we watch or listen to the weather forecast what we're most interested in is whether it's going to rain or not. In our bus-stop conversations the rain will feature more than anything else: either an observation of the rain currently falling or a pondering upon the probability – or most likely inevitability – of its arrival. We give it curious descriptions: there are stair rods coming down or it's raining cats and dogs. In Portugal it rains toads' beards, in Germany it pours cobblers' boys, in the Netherlands it rains pipe stems, and it's wheelbarrows in the Czech Republic, while tractors are falling in neighbouring Slovakia. In Denmark it rains shoemakers' apprentices and in Greece it rains chair legs. It rains female trolls in Norway while according to Afrikaans-speakers it's raining old women with knobkerries (apparently a type of shillelagh).

This range of descriptions from agricultural machinery to club-wielding pensioners is usually testament to the noise of the rain on the roofs of our dwellings, yet the sound of the rain usually has a soothing effect; a hissing sound that has an instantly relaxing ambience, at least if you're indoors. It could be some ancient genetic echo – if it's raining there's not much work that can be done out in the fields – or even further back: if it's raining the sharp-toothed, hungry predators out there are probably taking shelter the same as you are. That gentle, consistent noise like gas escaping from a thousand cylinders has an instant calming effect.

There is something mesmerising about the rain, whether it's watching it fall, listening to it or looking down at a puddle and seeing the expansion of an infinite number of splash circles. I can actually remember the first time I was truly spellbound by the rain. It was a pretty innocuous occasion too: I was in my early twenties, eking out an existence as a musician and on my way to visit someone who I hoped would prove to be the other half of a songwriting partnership that would make Lennon and McCartney look like hack jingle-writers. From memory I think he lived in Muswell Hill in north London, a place as far from my home in a gritty part of south-east London as it was possible to be. I remember walking out of the station into a wide, tree-lined street of huge, solid Victorian red-brick houses. It was early summer but the street was quiet. Having grown up on a main road and lived in places that were nowhere near as grand as this – where were the takeaways? The breakers' yards? The abandoned fridges and broken tellies on the pavement? – this was not the kind of road I walked down very often.

And the trees, the trees didn't look like they belonged on the street at all. They were tall, proud and thick-trunked, branches spreading wide, abundant with bright green leaves that almost touched their counterparts on the other side of the road, nothing like the thin, grey, pollution-choked specimens I was used to, protected by iron cages and held up by splints that seemed to be the only things preventing them falling into sickly swoons. I walked up the hill completely alone; there wasn't another soul about, just these grand, confident houses and the magnificent trees.

Then it started to rain, a warm summer shower making dark spots on the pavement that gradually joined to form a wet whole. The sound has stuck with me ever since, even though nearly two decades have passed. It was the steamy hiss of rain, but as it hit the trees above and around me it had a depth of sound that I'd never experienced before. I was used to the sound of rain hitting dustbin lids, awnings, binbags, a range of pitches and timbres, but this was a mass of sound: raindrops on leaves and tarmac. It was uniform yet paradoxically it possessed a depth I could barely comprehend. It was everywhere, all around me. If I turned my head the sound would come anew from the direction I faced: the same sound, yet different. And the smell, the smell was sweet and delicious, that sublime aroma released by the rain in the summer after a dry period that was given the name 'petrichor' in 1964 by two Australian scientists, called Bear and Thomas. It was they who had deduced that the smell of rain comes from the plant oils and proteins that have dried onto hard surfaces during periods of dry weather. When the rain hits them they release that unique aroma, petrichor, a word taken quite brilliantly from the Greek *petros*, meaning rock, and ichor, the substance that coursed through the veins of the gods. But it's the sound of the rain that has stayed with me to this day. I've heard it since, especially when out walking in rainy countryside, but that day was the first time I'd ever noticed the beauty in its sound. It suggested space, peace, exhilaration and security, all in the most perfect surround-sound. It was like being introduced to Beethoven, Elvis Presley and Louis Armstrong all at once, when you'd previously existed to a soundtrack of George Formby.

I'd like to say that this inspired my writing partner and me to uncharted heights of musical creativity, but it didn't. The success of the collaboration can be gauged by the fact I can't even remember his name. But that summer day in a strange part of London changed my appreciation of the rain for ever. Like the rain fossils of India and South Africa it was a long-forgotten rain shower on a long-forgotten day; I wouldn't even recognise that street if I saw it again, I couldn't even tell you the year it happened, but the sound and the smell of the rain lodged in my subconscious, instilling a sense of wonder that has never left.

Thinking back, it was the day that I fell in love with the rain. This was handy, because I had begun to realise the rain was head over heels in love with me.

3

RAIN: A LOVE STORY

I think the rain might be in love with me.

It certainly seems to follow me everywhere, hanging around unnoticed until I step outside and then greeting me with an emphatic shower of what I suspect is devoted adoration. That can be the only explanation for how I am rained on everywhere. I go to a football match and it rains. I walk to the post office and it rains. There's really no point in me going on holiday anywhere nice because it rains. Wherever I go the rain is there with me, caressing me, stroking me, smothering me in its soggy affection, rarely leaving me alone.

I try not to mind too much and certainly don't attempt to encourage it, but going out on a sunny day only to find that within minutes my hair's plastered to my head and there's a cold dampness seeping along my socks from the big toe can become a little wearing.

Looking back I can pretty much nail the day I realised that the rain had more than just a passing interest in me. For a short while after I left university I had an office job, as part of which I occasionally had to take documents to buildings in various parts of a large, leafy complex. It became a standing joke among my workmates that I would invariably come back from these errands soaking wet. Now there's nothing particularly remarkable about that: a wet day is a wet day and I was rarely organised enough to have some kind of waterproof jacket to hand, but these weren't always wet days. These were days that were bone dry save for the few minutes in which I was outdoors. On one particular summer's day the sun shone from what appeared to be a clear blue sky and, as I collected up the sheaves of reports and data ready to deliver them to one of the eggheads who analysed these sorts of things elsewhere in the complex, one of my colleagues nodded out of the window at the shadows pooled beneath the trees on sun-bleached pavements and said, 'Well, at least you won't get wet today.' I laughed, she laughed, the rest of the office laughed. I headed out of the office, pushed open the double doors at the entrance to the building, walked about twenty yards and felt the first raindrop ping off my cheek. Another thudded onto the top of my head. The temperature plummeted and tiny brown spots began to appear on the tarmac around me. The sky darkened suddenly and a familiar hissing noise grew in a dramatic crescendo. I pressed the document folders to my chest, hunched my shoulders and felt the coarse rubbing of my damp suit jacket against my neck. Every footstep sent a spout of water forward from the ends of my shoes and a familiar cold wetness began to spread back along my socks. My trousers soon stuck cold and clammy to my thighs and rivulets of water probed downwards inside my shirt collar. That day I would have told you exactly where you could poke your petrichor.

Ten minutes later, documents delivered just before the

cardboard folders disintegrated completely, I was pushing open the outer door of my building to stand dripping in the hallway. I brushed some of the excess water from my suit and shook my head like a sneezing terrier, launching a fan of second-hand raindrops that pattered onto carpet tiles and spattered against windows. My hair was saturated; my face was shiny, wet and cold. I opened the office door and walked slowly in with my shoes creaking and squelching and my socks squeezing out a swell of water with every step. The place fell completely silent and around twenty people stared at me, some with telephones to their ears in mid-conversation. Bosses came out of their offices to witness me squelch by and there wasn't a mouth closed in the room. Through the window I could see the sun beating down and already burning dry patches into the tarmac outside. I reached my desk, peeled off my jacket and hung it on a drawer handle. I pinched my thumbs and forefingers together and pulled my drenched shirt front away from my chest, then did the same to the shoulders. Nobody said a word and I caught nobody's eye; I just pulled the next pile of documents towards me, leaned forward and watched as half a dozen droplets fell on the paper, irrevocably blurring some of its inky information. By the time I dried out it was time to go home, and it rained all the way to the bus stop.

It's a double act that has been played out across the world, the rain and me. From the Isle of Wight to Shetland and from the west of Ireland to the flat plains of the Fens, from the Pacific paradise of Hawaii to sheltering beneath the awning of Sydney Opera House, from Memphis to Melbourne, the Isle of Man to the Isle of Dogs, wherever I've been in the world the rain has sought me and found me, adored me, caressed me, got beneath my clothes and against my skin. I watch it sweeping in from the sea and tapping enthusiastically against my window, or hovering over distant hills waiting for me to pop out for a paper and a pint of milk before rushing over to greet me.

It's not always a bad thing, mind. When I was in the school cricket team, I lived in fear of a particular fast bowler from another school, whose name was Cutler. Even my school's star cricketers would go pale and swallow hard when Cutler's name was mentioned. He was tall, rangy, had a run-up so long that it seemed to start in a different postcode, and an ice-cool demeanour that didn't express any emotion even when he sent a middle stump cartwheeling towards the pavilion.

Myths abounded about Cutler: the kids he'd put in hospital, the number of stumps he'd smashed to matchwood, even that he'd killed a boy one year with a ball that pitched just short of a length on middle-and-leg. None of this was true, of course, but the speed with which Cutler could propel a hard red ball along a cricket pitch didn't need embellishing, it was terrifying enough in reality. I'd begun fretting during the winter. I'd even written to the country's leading cricket coach about him: he replied graciously and helpfully but much of his letter seemed to revolve around getting my puny, quaking body and whimpering, crackable head in line with the ball, when the advice I'd been hoping for involved a more evasive kind of action, like emigrating.

Finally the day of the game dawned. I know it dawned because I saw the sun come up, having not slept a wink all night. The fact that it was clearly sunlight seeping around the curtains also meant the unseasonal blizzard for which I'd been fervently praying hadn't materialised either.

When Cutler's school arrived I looked at the boys disembarking from the minibus, hoping that he'd been struck down with a debilitating illness that caused his arms to fall off, but no, there he was, unfurling his unfeasibly lanky frame and stretching arms so long he practically qualified as an aircraft. They batted first, prolonging the agony, and then it was our turn. The usual hubbub of changing-room conversation was entirely absent as our opening batsmen donned what protective clothing was available: these were the days before helmets had filtered down from the professional game and thigh pads were the ostentatious cricket purchase *du jour*. Only a couple of our lads had them and there was already a waiting list to borrow them. You could have strapped mattresses to my front and back and plonked a motor-cycle helmet on my head and I'd still have been just as scared.

We trooped out and saw, in the far distance, the figure of Cutler at the end of his run-up, rubbing the ball casually against the front of his whites. The umpire called 'Play', and Cutler set off. A walk, a trundle, a jog, a sprint, an explosion of whippy arms and almost immediately the thwack of the ball in the wicket-keeper's gloves somewhere back near the boundary. In the changing room, somebody whimpered.

Wickets fell. I put on my pads and gloves. The moment I'd feared for so long would soon be upon me. The stumps were smashed and splayed again and again, but on the bright side no one was dead. Two more wickets and it would be my turn in the firing line. I was so scared I hadn't noticed it getting colder and the wind gusting up, ruffling my hair and tugging at my shirtsleeves: I'd just put the shivering down to sheer naked fear. Another wicket fell, meaning that there was just one raised umpire's finger between me and my Cutler destiny, but as the batsman before me made his reluctant way towards the wicket the sun went behind the edge of a cloud that hadn't been there a few minutes earlier, a dark shadow fell across the pitch and the raindrops began to fall. A few at first, then more. Then a steady downpour that had scorers, spectators and players alike sprinting for the shelter of the pavilion in a clatter of picnic chairs and a flurry of wind-whipped newspapers.

When puddles began to form on the outfield it was clear there would be no more play that day. I stood at the window, surveyed the soggy scene outside and laughed so hard I nearly did myself a mischief with my abdominal protector. The rain had, as far as I was concerned, saved my life.

At the start of the following term word spread that Cutler's dad had taken a job somewhere in Lancashire and the whole family had gone with him. I would never have to face him across twenty-two yards of cropped grass as long as I lived, and for that I had the rain to thank.

It took me years to realise the rain's apparent affection for me: holidays spent looking through windows at landscapes fish-eyed by raindrops; watching football matches among soaking crowds steaming gently under the floodlights; the heat of a pub fire as the rain that had thrown itself at me all that day evaporated from my walking boots. I was learning to love the rain back, recognising it as much more than a soggy inconvenience, seeing it as something more positive than a creator of puddles for buses to plough through just as I passed. Much as I'd become better disposed to the rain, however, I doubted I'd ever devote my life to it. No, to do something like that, and to do it selflessly and with mind-boggling levels of dedication, would take someone really extraordinary.

There is a photograph of George James Symons that must

have been taken towards the end of his life. Like many Victorians he has the kind of full, thick grey beard that looks as if it might actually be made of twigs, while his hair is swept back from his forehead. In many ways it's a typically stern portrait of its time, but the key to Symons can be seen in this photograph, in his eyes. There are pronounced bags under them, fatty half-moons picked out by the deep semi-circular shadows beneath, the legacy of long days and nights at his exhausting work. But it's the eyes themselves that grab you - even in this primitive black-and-white photograph, even though he's looking to the side of the camera. There's laughter in them, a playful bright-ness that suggests some kind of quip is forming, ready for the instant the picture has been taken. There's a perceptible bounce about him and energy bursts from the picture. His arms may be folded, his legs may be crossed, but this is clearly a man to whom sitting still didn't come naturally.

The beginnings of a smile can be detected through the dense thatch of his facial hair, but it's those eyes that strike you. They may be in monochrome but you can be pretty sure that they're a bright, sparkling blue, the colour of raindrops in sunshine on a window pane. Implausible as it sounds, you can just tell that he was a lovely man, the kind of man of whom it would be written in his obituary that 'he attained the rare distinction of passing through life without making a single enemy'. He looks like just the kind of guy you'd be delighted to sit at the bar with over a couple of pints, only in his case the chances are he wouldn't have the time. Genial he may have been, but George James Symons had a singular obsession. It was one that mani-fested itself at an early age, one that caused him at least one nervous breakdown and one that kept him working solidly right

up until his death. It was an obsession that prompted him to
resign from his one and only proper job in his twenties and
devote himself to it unswervingly for the rest of his life. Symons's
obsession was rain. He was obsessed, but not in a way that would
have him cavorting in a downpour wearing only his undershirt
and singing a *hey nonny nonny*; no, Symons's obsession was much
more practical and infinitely more useful: he dedicated his life
to the collection and interpretation of rainfall data.

Born in London in 1838, five weeks after Queen Victoria's
coronation, it's possible that the heavy droughts in England
during the 1850s might have triggered Symons's fascination
with precipitation. Either way, by the age of seventeen the
young George – 'a shy lad with a shock of red hair and a nervous laugh', according to a contemporary – was not only
presenting papers at the Royal Meteorological Society but being
made a Fellow of the organisation.

In 1860 Symons was studying to be a science teacher when
his physics lecturer John Tyndall, scientific adviser to the Board
of Trade, recommended him for a job in the Board's fledgling
meteorological department. Symons started collecting the data
with a scrupulous zeal that unnerved even the ultra-diligent
Robert FitzRoy, the department's head. Within months of
starting work Symons published the first volume of *British
Rainfall*, in which he'd collected and assimilated data from 163
locations around Britain. Frustrated by the gaps in the data that
prevented the appreciation of a fuller picture, he set about
recruiting volunteers and over the next forty years he would
increase the number of locations from 163 to 3528 by the sheer
force of his personality. Such was his zeal for collecting rainfall
data that, after FitzRoy pointed out he was neglecting his other

departmental duties at the Board of Trade, Symons resigned and concentrated on his passion full time, 'unpleasant as it was financially', as he himself admitted.

Despite the lack of resources Symons threw himself into his work, maintaining his ever-increasing band of volunteers by post and by occasional tours of the country to inspect the rain gauges in which they were collecting their figures, often with mixed results. Not least because there was no standard type of gauge, nor any guidelines for their positioning. One man in the Lake District recorded how Symons visited him in 1866 and 'cruised for hours among the rocks . . . in search of my gauges. He could not find one of them.'

Symons's agreeable personality and epidemically infectious enthusiasm meant that complete strangers right across Britain were meticulously noting data from their often home-made rain gauges and sending them to Symons's house at 62 Camden Square in London, where he would interpret them for publication in his annual. His efforts were heroic but took their toll on his health: in 1864, soon after leaving the Board of Trade, it seems he suffered some kind of breakdown and couldn't work for nine months (at one stage there was some doubt as to whether he'd ever be able to work again), and something similar happened in the mid-1870s. Yet his passion and enthusiasm never dimmed, to the extent of searching out historical weather records, appealing for anyone who'd kept their own rainfall records prior to 1860 to send them his way. He was phenomenally successful even at this – it is thought that he gathered some seven thousand sets of historical records that enabled him to compile reliable national rainfall records as far back as 1815 (one box of journals that came into his possession was a remarkable unbroken series of records

kept by Thomas Barker, a squire from Lyndon in Rutland, that
ran for almost six decades between 1736 and 1794).

Symons's was a labour of love. His income came from the
occasional grant, but mostly through subscriptions to his British
Rainfall Organisation and *Symons's Monthly Meteorological
Magazine*, which he somehow found the time to compile and
publish. His financial existence was highly precarious but, even
as a man of high intelligence and capabilities who could have
walked into a very good job, his dedication to the rain never
wavered as long he lived. His was an almost superhuman passion
for something that, its importance to crops and reservoirs aside,
most people regarded as an inconvenience at best. It's thanks to
Symons's tireless efforts and those of his volunteers that we
understand so much about our rainfall, its patterns, its quirks, its
excesses and its failings. It is a subject close to our hearts and it's
largely down to this genial single-minded Victorian workaholic
that we know as much as we do.

Symons was beginning his fifth decade of devotion when his
life of relentless toil caught up with him. In February 1900 he
suffered a stroke that left him paralysed and unable to speak, a
particularly cruel blow to a man whose life was so active, rig-
orous and dependent on communication. I visited Camden
Square and stood outside his house, with its plaque commemorating his time there, and looked up at the windows, wondering
if one had been that of the bedroom in which he'd lingered for
four weeks, able to do nothing more than listen to the patter of
raindrops. I hoped that his fear of the stroke's ultimate outcome
and intense frustration at being unable to work might have been
tempered a little by the thrill of hearing the sound that had
first inspired him on his lifelong quest, because Symons was no

mere statistician: he genuinely loved the rain. Such was the meticulous nature of his work that even in his magazine writings Symons revealed little about the nature of his passion, but one contemporary reported finding him one day in the middle of a thunderstorm sitting at the window while holding a barometer almost in a lover's embrace, gazing out at the maelstrom with a look of wonder in his eyes and a faraway smile on his face.

While his life was dedicated to the collation, interpretation and publication of rainfall statistics, I like to think that it was the sound of the rain that had first inspired Symons's mission. Maybe as a youngster after weeks of fearsome drought some rain had started to fall, speckling the window to which the young George rushed and looked out as the Pimlico pavement turned from grey to shiny brown and he could sense the parched, heavy air being rinsed through and cleansed of the dust and dirt that had hung there for weeks. All those years later at the dawn of a new century he might well have heard it again. His wife had predeceased him and their only child had died in infancy many years earlier. In that sense he was alone at the end. But in another, there was the rain and the knowledge that in the garden his collection of rain gauges was still out there, gently filling with water ready for readings that would still be published long after his death on 10 March 1900.

It is perhaps appropriate that Symons died just a couple of months into the twentieth century because he was definitely a man of the nineteenth. With his drive, his diligence, his inflexible dedication to one subject, his far-sightedness and his embrace of innovation he was utterly Victorian. It's likely you've never heard of George James Symons, but for me he is unquestionably one of the great Britons. He combined a fearsome intellect, extraordinary dedication and a tangible streak of eccentricity to devote his life to the recording of what must have seemed – and still seems – to most people a mundane irritant, but the legacy of his work remains in every weather forecast in every medium. If it hadn't been for Symons our understanding of the weather would be much, much poorer. Our ability to forecast the weather would have much less depth than it does today; he discerned patterns of rainfall that have helped generations of farmers with their crops, and it is thanks to Symons that when we hear of the highest (or lowest) rainfall levels 'since records began', we know those records go back as far as they possibly can because of the tireless work of George James Symons.

He is buried in Kensal Green cemetery in north London and – a thoughtful touch – his headstone was made from slate mined at a quarry near Seathwaite in the Lake District; a place that Symons would almost certainly have visited and one that would have been very close to his heart.

Eight miles south-west of Keswick, in a quiet valley at the very end of a minor road off the Honister Pass, is a small farming hamlet. There are perhaps half a dozen whitewashed terraced cottages and a few barns, and on the face of it it's an unremarkable spot: typically breathtaking Lake District scenery notwithstanding, there's very little in Seathwaite to detain you. Despite this, Seathwaite holds a very special distinction, a very British claim to fame that arises from a small gravel area enclosed by wooden fencing, which contains what look like a couple of large plastic syringes sticking out of the ground.

They're rain gauges; the rain gauges that confirm with every reading that Seathwaite is the rainiest inhabited place in Britain.

This tiny hamlet on a plateau between mountains records around 140 inches of rainfall every year. That's a lot of rain. Consider that Manchester, a city with a reputation for precipitation like no other, gets around forty-four inches every year and London just twenty-five, you can see just how wet Seathwaite really is.

In September 1966 the Seathwaite gauges recorded five inches of rain in an hour, as part of a downpour that caused widespread flooding in the area. Seathwaite was at the forefront of rainy records again in November 2009, when an astonishing 12.38 inches of rain fell in a twenty-four-hour period. That's nearly half of London's annual rainfall hosing down in the space of a single day. A deep depression had come in over the Atlantic and passed over Ireland, causing storms across the country as well as flooding, then moved across the Irish Sea to arrive in Cumbria on 19 November. Cockermouth was flooded to a depth of more than eight feet, and so sudden was the deluge that two hundred people had to be rescued from their homes. A policeman died when a bridge over the River Derwent on which he was diverting traffic collapsed beneath him and he was swept away.

So why is Seathwaite so soggy? Cumbria is wetter than most places because of its topography: it's England's most mountainous region. Warm air flows in from the Atlantic and is pushed sharply upwards to form clouds that then fall back to earth as rain. Mainly, it seems, over a tiny cluster of cottages and barns in Borrowdale. Hence it was only appropriate that George James Symons's headstone should have come from a quarry so close to this sacred ground of British rainfall.

Incidentally, we have Seathwaite weather to thank for one of the greatest everyday inventions in history. In the middle of the sixteenth century a storm blew over a large ash tree not far from the hamlet, after which local shepherds noticed globules of a shiny black substance tangled among the roots. On further investigation they found that the lumps left a black mark on everything they touched, a mark that wasn't washed away by the rain. The shepherds began to use it to mark their sheep, rubbing the substance against their fleece to provide a weatherproof way of identifying their flocks. The downside was that it stained their hands too, so the shepherds began encasing it in string or wool. The shiny black substance was graphite and the shepherds had, if the story is to be believed, just invented the pencil.

By chance the shepherds had discovered what proved to be the purest, most solid source of graphite that would ever be found. Its appearance led scientists at first to mistakenly classify it as a type of lead – hence to this day we still refer to pencils as being made from lead – and it was originally named 'plumbago', the Latin for 'lead ore'. In addition to its usefulness in sheep identification, graphite was found to be a perfect lining for the moulds used in making cannonballs and the Seathwaite graphite seams were soon requisitioned by the army, with an armed guard placed permanently at the entrance to the mine workings. Eventually the more peaceful pencil became graphite's main product and led in 1832 to the formation of the Cumberland Pencil Company in Keswick, a company that makes artists' pencils to this day and, rather brilliantly, is also home to a pencil museum that, among other things, boasts the world's biggest pencil (26 feet long, since you ask).

Pencils aside, Seathwaite is an exceptionally soggy place. It is, however, positively arid compared to the wettest place in the world and, appropriately for somewhere as defined by rain as we think *we* are, it has a story that could be an Ealing comedy.

Cherrapunji and Mawsynram are small towns five thousand feet up in the Khasi Hills. Yes, the Khasi Hills. The hills are in the Meghalaya region in the far north-eastern corner of India, where the two towns squabble over which has the right to the title of the rainiest place in the world. Cherrapunji has long laid claim to the crown, ever since the British East India Company made the town its operations base for north-east India in the early 1850s and set up a weather station. During the monsoon season it rains literally like nowhere else on earth, hammering down for days and days on end without respite.

However, Cherrapunji's fame as the wettest place on earth – the weather station there records an average annual rainfall of 463 inches, roughly twenty times what London receives – irked the inhabitants of Mawsynram, a dozen miles away, who were convinced their rainfall was higher. Mawsynram, whose inhabitants still manufacture and wear tartan shawls as a legacy of the Scottish missionaries who established themselves in the area in the early nineteenth century, takes its own readings which, over a year, do seem to be slightly higher than those taken in Cherrapunji: 467 inches to the latter's 463. However, the people of Cherrapunji point out that Mawsynram has no permanent weather station: its readings are taken by a local government official as a small part of his overall duties. When he's unavailable the recordings are usually taken by a local chef. As far as the Cherrapunji weather observers are concerned this is not good enough to eclipse their official readings, which are taken meticulously twice a day and filed with the Indian meteorological office.

It may not be much consolation to the people of Mawsynram, but they probably get to enjoy some outstanding rainbows. One of nature's most spectacular tricks, the rainbow is a lightshow of breathtaking beauty and perfection. If anyone doesn't exclaim, or at least catch their breath, every time they see a rainbow, well, they're probably not human. There is something about the rainbow that makes us just feel better about things. It makes us want to share it too: we'll point it out to companions and even strangers. We'll coo and sigh, taking a moment from the stresses and pressures of our day to bathe in the calm of the rainbow. With rainbows there's a definite sense that everything's going to be all right – it's like a new meteorological dawn, the rain's gone away and the sun's coming out and in the meantime here's this magical swathe of colour in the sky.

When I was ten years old we went on a family holiday to Hawaii. I did some tremendous ten-year-old's things: ate enormous ice-cream sundaes, had pancakes for breakfast, saw places I recognised from re-runs of *Hawaii Five-0* and drank my own body weight in pineapple juice from a special drinking fountain at the end of a pineapple cannery tour. I practically had to be rolled back to the coach. But the thing I remember most about that holiday, the image I can still see as clearly as if it were yesterday, is the day we saw the end of the rainbow. We were on a bus to Pearl Harbor, and as the bay came into view there was a perfectly defined rainbow with one end disappearing into the sea. It seemed so close that I could almost take a running jump and slide down it into the water, but clearly time has brought the rainbow much nearer to me than it actually was: nobody gets close to a rainbow. However, the awe induced by the sight of that wonderful, sharp-edged, solid-looking rainbow has never left me and I hope it never will.

The magic of the rainbow is in the fact that it doesn't actually exist, at least not as a physical entity. The rainbow isn't there, it appears only in your eye. You can't go under a rainbow, you can't go round it. You can't lean against it and light a cigarette because it's not actually there. Move towards it and it'll back away or even vanish altogether.

To see a rainbow you need to have the sun behind you at a fairly low angle and, most commonly, raindrops in the sky ahead. Ideally there will also be dark clouds ahead, to really help give the rainbow definition. Of course it doesn't have to be rain – rainbows can be seen in the spray of fountains, or the mist created by a waterfall – but it's rainbows that are actually in the rain, the horizon-straddling bars of symmetrical colour, that we love most.

Explaining the rainbow feels a little like letting daylight in upon magic, but it's one of nature's most wonderful processes. The sun's rays are refracted as they hit the surface of the raindrop, then they're reflected off the back of the drop before being refracted once more as they leave the raindrop, again through the front of the drop. This process splits the light into what most of us see as a spectrum of seven colours as it hits our eyes: red, orange, yellow, green, blue, indigo and violet – what we see as the rainbow.

While it was Isaac Newton who most accurately defined the rainbow for us, some of the world's greatest minds had tried to explain it. Seneca wrote at length about them in *Naturales Quaestiones*, and he wasn't far off explaining them correctly either. A range of thinkers across the world, the likes of the eleventh-century Chinese scholar Shen Kuo and the thirteenth-century Persian philosopher Qutb al-Din al-Shirazi, for example, concluded that it was something to do with the sun's rays hitting water droplets. René Descartes came closest to correctly identifying the nature of refraction, but it took Isaac Newton refracting light through a glass prism to finally reveal the science behind the rainbow (much to Keats's later chagrin: in his 1820 poem 'Lamia' he complains that 'Philosophy will clip an Angel's wings, /Conquer all mysteries by rule and line, /Empty the haunted air, and gnomed mine – /Unweave a rainbow ...).

Thankfully, knowing the secrets of the rainbow doesn't lessen our delight when we see one. We regard it with the same awe that caused the Vikings to conclude the rainbow was a bridge between our world and that of the gods, and the ancient Greeks and Romans to presume it was a path used by the messengers of the gods to travel between the heavens and the earth. This view of the rainbow as some kind of transition is entirely under-standable as it's arguably the most magical, beautiful and enigmatic aspect of the weather: impossible to pin down, capture or touch and utterly breathtaking to look at. It can't have been wholly of this world.

Wordsworth wrote of how

My heart leaps up when I behold
A rainbow in the sky:
So was it when my life began,
So is it now I am a man,
So be it when I shall grow old
Or let me die!

The rainbow is nature as artist; an impressionist's vision of peace and calm now the storm has passed. The rainbow almost has the flourish of nature's signature. It is certainly in stark con-trast to the other overtly spectacular aspect of the weather – lightning.

4

LIGHTNING: NATURE'S BLING

We never get tired of lightning. We might not like it very much, but everyone feels the same faint thrill when there's a flash outside the window. Like meerkats, our heads pop up, and we ask, 'Was that lightning?', half hoping it was, half hoping it wasn't. We keep still and quiet for a few seconds, waiting for the rumble that confirms that yes, it was definitely lightning.

Sometimes we catch sight of it by chance, a sudden fork in the sky that's gone as soon as we look: it's always a slightly illicit thrill to see lightning. And is there any greater, more stomach-churning feeling of suspense than when counting the seconds between the flash of lightning and the start of the low rumble of thunder to gauge whether the storm is getting closer?

Lightning is meteorology's bling. It's Mother Nature showing off; the weather's flashy guitar solo. It's spectacular, sublime and beautiful, and must have terrified and fascinated us since the dawn of humanity. There would have been massive lightning storms at the earth's very formation: the jagged silver-white forks we see in the sky today are the oldest show on earth, performing constantly since the beginning of time. Yet it's only in the last 250 years or so that we've learned how to harness and imitate lightning, for what is electricity but domesticated lightning? Or, as George Carlin would have it, 'organised lightning'.

Benjamin Franklin gets all the credit for recognising and harnessing the power of lightning but, while he certainly deserves most of it, there are a couple of other names that should be mentioned in dispatches. It's a wonder that Franklin had time to mess about with lightning at all, what with being a politician, diplomat, writer, printer, musician and postmaster, not to mention inventing bifocals and daylight saving time, being one of the founding fathers of the United States, opening America's first public lending library and developing the musical instrument known as the glass harmonica (which is all well and good, but how many beer mats could he flip and catch? My game, I think).

When he noticed clouds gathering over Philadelphia on 15 June 1752 Franklin could wait no longer. He'd long suspected that lightning flashes were electrical; indeed, three years earlier he had listed twelve similarities between the static electricity he created in his laboratory and lightning, and was waiting for the proposed sixty-foot spire (destined to be America's tallest man-made structure of its time) of Christ Church to be completed so he could affix a rod to its pinnacle and test his theory. He'd written to scientific friends in Europe outlining his ideas and suggesting that they might be tested by use of an iron rod forty feet high, which 'when such clouds are passing low might be electrified and afford sparks'. It's not really surprising, then, that when the storm clouds began to churn and the wind to blow harder that summer day in 1752 he decided to take matters into his own hands rather than wait for the builders to finish the steeple. He came up with the idea of using a kite, tying a key at the end of the string, which in turn would be attached to a bollard by means of a piece of silk. Suitably equipped he headed out into the storm, set up his apparatus and launched the kite.

Today, of course, we know that flying a kite in a storm is the height of insanity. It's like Newton deciding that the apple wasn't conclusive enough and further investigating his theory of gravity by having someone drop a bowling ball on his head from a second-floor balcony. It can be argued that Franklin would have had no idea of the extent of the extraordinary power about to be unleashed in the heavens, but he must at least have known that people were regularly killed by lightning strikes. As electrical experiments go, this was about as far from rubbing a balloon on your jumper, sticking it to your hair and going 'Ta-daaa!' as you could get. But he was Benjamin Franklin, a man who always knew what he was doing.

As the storm drew nearer Franklin noticed that the fibres on the rain-soaked kite string were standing erect. He then placed his hand close to the key, which caused a bright spark to leap from the metal to his flesh, like reaching for a brass doorknob when standing on a carpet in stockinged feet on a cold day. His hypothesis, he realised, was correct. Presumably at this point he wound in the kite before the storm really hit, but as he ran through the rain to his house, reached the front door, rummaged in his pockets, swore under his breath, sprinted back to the bollard, untied the key, ran back to the house again and let himself in, Franklin knew he was really onto something. But, as it turned out – and as Franklin himself was one of the first to acknowledge – he hadn't been the first. Thomas-François

Dalibard, a French scientist with whom Franklin had corresponded, had actually carried out the forty-foot-pole experiment the previous month, but news of it had not reached Philadelphia by the time the clouds began to gather and Franklin hauled the kite out from the cupboard under the stairs. It was the American who would earn the plaudits, and in fairness it seems his experiment was more conclusive.

Less than a year after Franklin diced with death, lightning-based electrical experimentation claimed its first victim. Georg Richmann was a German physicist who in 1741 had been elected to the St Petersburg Academy of Sciences on the strength of his research into electricity. He'd heard of Franklin's iron-rod conductor theory and, based on Franklin's idea, set up a contraption of his own devising in his laboratory. He attached a wire to the top of his house, which was in turn connected to a suspended iron bar that hung over a bowl of iron filings. He'd also set up a needle and gauge to act as an indicator of what was going on during a storm.

On 6 August 1753 Richmann was at a meeting at the Academy when he heard the distant rumble of thunder. Realising that this could be the first chance to test his new apparatus he headed home as fast as he could, taking the Society's engraver Mikhail Sokolow with him to record the scene for posterity. The storm arrived soon after they reached the house and an excited Richmann led Sokolow to his laboratory. The apparatus was in place and the conditions were right: a thunderstorm was now raging directly overhead. An apprehensive Sokolow hesitated by the door but Richmann walked straight in, assuring him it was all perfectly safe. He'd just bent down to check the position of the needle when, according to Sokolow, a bright ball of lightning – described in the Royal Society's records as 'a globe of blue fire as large as [a] fist' – emerged from the iron bar, detached itself, floated towards Richmann and, on touching the scientist's head, caused a terrific explosion. Sokolow was knocked out cold and when he came round Richmann was dead, his clothes were singed and smoking and his shoes had been blown clean off his feet. The door was hanging off its hinges and the door frame in which Sokolow stood was split all the way round.

What Franklin and Richmann wouldn't have known – but Sokolow now had a pretty good idea – is just how fearsome lightning can be. When lightning discharges it can contain a billion volts, travel as fast as 140,000mph and reach temperatures of up to 30,000°C (it's that superfast heating, causing an explosive expansion of air that makes thunder). It is an incredible amount of energy; even more so when you consider that in the time you'll take to read this single page there will have been roughly six thousand lightning strikes around the world.

Despite this wealth of statistics from the very heart of the flash, science is still not entirely sure what causes lightning. One thing you do need is a cumulus cloud that can raise moisture to altitudes of three miles or more, where the air temperature is below freezing. Once ice crystals begin to form they become cumulonimbi – storm clouds. One theory has it that as ice crystals begin to form in the freezing air they bump into each other and into water droplets that are still unfrozen. These tiny collisions produce an electrical charge which, when spread over a wide area inside the cloud, combine to produce the massive discharge of lightning. There are two basic types of lightning: intracloud, which stays within the cumulonimbus and appears as a flash of light like a sheet in the sky, and cloud-to-ground, the really spectacular one that occurs when a leader of electrons drops from a storm and meets a positive charge coming the other way from the ground, producing the visible flash. Its frequently jagged form is because each leader pauses for a fraction of a second on its way down, changing direction and sometimes splitting into various branches as it looks for the easiest route to earth. It's this mazy itinerary being lit up by the positive charge that gives lightning its forks and crooked appearance.

Lightning looks for the quickest and easiest way to earth. Fortunately fewer people are struck by lightning these days than ever before, which is especially reassuring when you consider the surprisingly high incidence of lightning in Britain and Ireland. Believe it or not, on average each square kilometre of ground in England and Wales is hit by lightning twice a year, which sounds scarily high to me.

Don't worry, though: your chances of actually being struck are pretty slim. Thirteen-million-to-one slim, in fact, a level of probability that has stayed steady since around 1960. In the forty years prior to that the figure was four million to one, while in the late nineteenth century the odds were an almost nailed-on cert at a paltry 1.6 million to one. It's a wonder anyone ever stepped outside. The reason for these lengthening odds is, quite simply, that we don't work outdoors as much as we used to. Farm labourers were the most at risk by far, while the number of fatalities among church-bell ringers was disproportionately high before the Enlightenment. This was because in those more superstitious times thunderstorms were interpreted as God's wrath. Many people would shelter in church and pray for God's forgiveness for whatever the community had done to incur the violence of His disapproval and the bells would be rung, partly to tell the Almighty that the people were really, really sorry and also because many believed that the storm could be persuaded to leave by a loud peal of church bells. These were the days before lightning conductors were in widespread use and as lightning looks for the best way down it's usually the tallest building in the vicinity that's struck. More often than not this meant the one with the steeple, making churches probably the most dangerous places in which people could gather. The poor old bell ringers, their damp and sweaty ropes the perfect conductors, were usually the ones who ended up fried.

The divine aspect of lightning has been around ever since we first started worshipping things. Zeus, the Ancient Greeks' most revered god, was the god of thunder and he liked to carry a thunderbolt around with him. Anywhere that was struck by lightning was hence immediately designated a sacred place. Indeed, the continued presence of the Pope in Rome is partly down to lightning. During the Avignon papacy, the English, Spanish and Germans objected to paying their tithes to a French-based Church. The school of cardinals was summoned to a vote on whether to stay in Avignon or go back to Rome. Given the number of French cardinals that had been appointed in case just such an eventuality arose, Avignon stood a fair-to-middling chance of hosting the papacy for as long as it pleased. Until, that is, a bolt of lightning struck the chamber in which the discussions were being held. The cardinals interpreted this as an expression of divine displeasure and before long the papacy was on its way back to the Vatican with just some very good cheese and a basket of baguettes to show for it.

Nearly six hundred years later, in 1987, the Pope was in Miami, addressing a crowd estimated at somewhere north of three hundred thousand, when lightning intervened again. His Holiness was holding forth on the usual laugh-a-minute subjects like contraception and sin when suddenly there was a bright flash, a bang, and a bolt of lightning hit the ground a few yards in front of him. In 1984 York Minster was struck and suffered major damage just three days after Dr David Jenkins had been consecrated there as the Bishop of Durham, an appointment that had been the subject of some controversy given his apparent view that the Resurrection was 'a conjuring trick with bones'.

Despite the fact that the odds are generally in our favour when it comes to the chances of being struck by lightning – even if we have apparently incurred the wrath of a deity – there have been some notable exceptions. Take a British Army major by the name of Summerford, for example, who at the tail end of the First World War was struck by lightning and thrown from his horse in Flanders, and temporarily paralysed from the waist down. When he had recovered and with the war over, Summerford emigrated to Canada where, while out fishing in 1924, he was struck by lightning again. This time his whole right side was paralysed and it took two years for movement to return. Six years later he was struck a third time, and on this occasion the paralysis turned out to be permanent. The poor major lingered until 1932 before handing in his pail. Two years later the cemetery in which he was buried was hit by a lightning bolt during a storm, and guess whose tombstone took the full force of the impact, being smashed to pieces in the process? (Clue: Major Summerford's.)

Being struck four times by lightning is pretty impressive, not least when one of those occasions is posthumous. However, the global award for being a human lightning conductor goes to a

Virginia park ranger named Roy Sullivan, who during his lifetime was struck no fewer than seven times.

The first occasion was in 1942. Sullivan was on duty in the
Shenandoah National Park when a storm blew up and he took
shelter in a new lookout tower. Alas for Roy, the lookout tower
was so new its lightning conductors hadn't been fitted yet, which
meant the storm looked at the ranger like a Mexican bandit
looks at a captured gringo before firing at his feet to make him
dance. Sullivan estimated the tower was hit seven or eight times
while he was in there so, realising he was in just about the worst
place he could be, he decided to make a run for it. He'd got just
a few yards when a bolt of lightning hit him, searing a half-inch
wide stripe down his right side, blowing off the big toenail of his
right foot and burning a hole right through his shoe.

Twenty-seven lightning-free years passed, then Sullivan commenced a remarkable run of being struck six times in eight
years. The 1969 strike was probably the unluckiest of the lot.
He was in his truck, probably the safest place he could be, when
a bolt of lightning bounced off a tree and shot through the open
window of his driver's door. This strike cost him his eyebrows
and much of his hair, but could have been a lot worse: he was
knocked unconscious and the truck rolled to halt just a few feet
from the edge of a cliff. A year later Sullivan was the victim of
another ricochet: he'd finished his shift and was walking up to
his front door when a lightning bolt hit an electrical transformer
box on a nearby telegraph pole, bounced off and sheared a deep
burn into the unfortunate ranger's right shoulder. Two years
later, in 1972, he was somehow hit *inside a ranger station*, and the
strike set his hair alight. He ran to the sink but the taps were too
low to get his head under and he was forced to use wet paper

towels instead. It was after this fourth strike that Sullivan realised something out of the ordinary was going on, that for some reason the weather was out to get him. By all accounts he started to become gloomy about his own mortality around this time and, on a more practical level, took to keeping a bucket of water in the back of his truck.

The following year Sullivan saw a storm approaching while he was driving and tried to outrun it. When he thought he was safe he parked his truck, opened the door and *pow!*, the lightning hit him as soon as his feet touched the ground. This strike again set his hair on fire and also blew off his shoes. Finally in 1977, when he was out fishing a bolt of lightning hit him right on the top of the head, leaving bad burns on his chest and stomach. A local newspaper report showed him posing self-consciously for a photograph, holding up his fishing hat to show a big oval hole in its crown.

You could say Sullivan was unlucky, but then he was struck by lightning seven times and yet survived. Lightning strikes have been known to change people, physically and mentally. There was a sixteenth-century priest in France who was shocked out of a lifelong paralysis, and in 1994 an Oklahoma woman named Mary Clamser was apparently cured of multiple sclerosis when lightning struck her house while she was touching a metal pole and wearing her metal leg braces. (Not only that: the following year she had to cancel an appointment at the Oklahoma City federal building, as she was going to California to talk about her experience. The appointment had been for nine o'clock on the morning of 19 April 1995, pretty much the exact moment the building was destroyed by Timothy McVeigh's truck bomb.) In 1971 Edwin Robinson of Falmouth, Maine, was involved in a bad car accident that cost him his sight and his hearing. Nine years later he was sheltering from the rain under a tree when it was struck by lightning, knocking him unconscious. He came round, groggily made his way home, went to bed and the next morning woke up to find he could see again. A day later his hearing returned. These are just individual strikes, but *seven*? Including six in eight years? Roy Sullivan's job meant that he was outdoors much of the time but even so, the odds of being struck once would still have been more than a million to one. If he did become gloomier after the fourth strike who could blame him?

The fishing-trip strike (in the immediate aftermath of which Sullivan also had to deal with a bear that had come looking for his catch) proved to be his last run-in with the weather gods, but the Roy Sullivan story does not, alas, have a happy ending. In 1983, at the age of seventy-one, he shot himself in the chest, by all accounts the result of unrequited love. The French have a phrase for falling in love at first sight: *un coup de foudre* – a lightning strike. The circumstances of Roy Sullivan's death add a tragic irony to the end of a man who had been struck by lightning seven times yet still was unlucky in love to the point where he felt he couldn't carry on. He was a pretty taciturn man who, on the face of it, seemed to take his lightning magnetism in his stride, but with each strike he must have wondered whether there could possibly be another and whether he would survive it. We all look up apprehensively as storm clouds gather and feel a certain quickening of the pulse whenever we see that brilliant white flash, but imagine if there was the very real chance that nature's most powerful weapon might be heading for you yet again! Not only that, but wondering whether the next thunderstorm might be the one that actually kills you. Thinking about it, Roy Sullivan was probably one of few people in the world who would have been delighted to hear thunder: if he could hear it, it meant he was still alive.

Before leaving lightning and its consequences, there is the tale of one person who, according to legend, was not delighted to hear thunder. John Dennis was a dramatist and critic of the late seventeenth and early eighteenth centuries and is allegedly responsible for one of our most common weather-related expressions, though he wouldn't have thanked you for mentioning it.

A man of independent means, Dennis was able to devote most of his time to creating and critiquing literature, but as a playwright he had mixed fortunes. His 1704 work *Liberty Asserted* was a roaring success, but this may have been due to its merciless anti-French sentiment – always a winner with English audiences in the early eighteenth century – because he struggled to match it for the rest of his career. He lived out his later life in poverty and died early in 1734, three weeks after a benefit was held for him at the Haymarket Theatre. It's for an incident in 1709 that he's remembered, however. His play of that year, *Appius and Virginia*, had not been well received, closing at the Drury Lane Theatre within days. The play would have been forgotten altogether were it not for the machine Dennis had devised to create the effect of thunder at appropriate moments during the play. When the sound effects garner more attention than the play itself, it's probably a sign that it's not going to be a winner.

A few months later Dennis was in the same theatre, apparently watching *Macbeth*, when he heard the very same thunder effect he'd created for his own play. 'By God,' he cried from the stalls, 'they will not let my play run but they will steal my thunder!'

★

Despite the affection that rain has always shown me I've had only one close encounter with lightning. It was very early on a Saturday morning and I was not in the best of form. I was on a bus, on my way to a literary event, I didn't really know where I was going and, although it was the height of summer, it was chucking it down. This was no refreshing shower that would rinse the muggy air clean, this was proper, heavy rain. I'd got soaked on the way to the bus stop and the prospect of hanging around in a rainy field for the rest of the day didn't exactly have me skipping a hot cha-cha. I was sitting at the front of the top deck and the bus was wheezing its way through parts of town I didn't recognise towards a soulless hotel by a roundabout on a ring road where I'd join my fellow speakers on a 'writers' bus' out to the festival. I was in a curmudgeonly mood anyway and the prospect of travelling on something that had actually been deliberately designated a 'writers' bus' didn't improve it. I wasn't even sure why I'd been asked. I had a book to plug, sure, but the rest of the bill seemed to be highfalutin novelists and playwrights whose work I'd either never read or, if I had, didn't particularly like. I had already decided that they were just the sort of people who would approve of such a thing as a 'writers' bus' and indeed relish the thought of travelling on one. I was picturing something like the teams travel in at World Cups, their country's name painted along the side and some kind of vacuous improving slogan underneath. I pictured a gleaming coach with The Writers' Bus written in giant letters along each side in copperplate script and an illustration of a quill clasped in a hand whose wrist was encased in a Shakespearean lace cuff.

I had little or no idea of where I was – all I knew was the

name of the hotel and the number of the bus on which I was sitting. I'd got wet while waiting for the bus and the copy of the book from which I was due to read on the stage had also got wet, the edges of the pages now swollen and crinkly in my coat pocket. Between the rain and the steamed-up windows I could barely see out, let alone work out where in the world I might be. I wiped the window next to me with my sleeve and saw that we were travelling on a road alongside a dual carriageway.

I hunched down in my seat and rested my head against the cold window, and was just deciding which chapter I should read when the lightning hit the bus.

From nowhere there was a blinding white flash and a bang so loud it seemed to come from inside my own head. It was so loud, in fact, that I seemed to absorb it, not just via my ears but through my whole being, before it zoned in on the deepest part of my brain. The bus skidded to a halt with a honk of protesting brakes and the engine stalled to silence. The echo of the explosion died and I realised that it was accompanied by the fading echo of a very loud swearword that could only have come from me. Red things like tiny worms swam in front of my eyes and the ringing noise in my ears gave way to the hiss of rain and tyres on wet tarmac from the dual carriageway.

The driver came up the stairs. 'Everyone OK?' he asked. The three or four of us scattered along the top deck nodded, our mouths hanging open.

'Jesus, I think we were hit by lightning there,' he said. 'Either us or that.' He nodded out of the window at a telegraph pole next to the bus.

'No, I think it was us,' said a man halfway back, 'the flash I saw was right next to that guy at the front.'

He meant me. It seemed I'd been inches away from where the most violent of the world's atmospheric discharges had used our bus to send about a billion volts to the earth. When my ears had stopped ringing and the wriggly red things in my eyes had swum off I was almost disappointed to note that my hair wasn't standing on end, my clothes weren't singed and frayed at the edges and I still had my shoes on.

The driver went back downstairs and restarted the engine, crunched the bus into gear and we pulled away. The whole thing suddenly felt anticlimactic. We'd just been struck by lightning, for goodness' sake. Yet people were going back to their newspapers or looking glassily out of the window while a '*tss, tss, tss, tss*' leaked from their earphones: it was almost as if a light bulb had just popped rather than a bolt of lightning had hit us. We'd just been on the receiving end of nature's most spectacular trick, but now we were just a bus again, like all the other buses.

We rounded a corner and there was the hotel. I got off. The writers' bus was just a bus like any other, the only calligraphy on the sides was the name of the company that owned it and the other writers were all perfectly nice. The rain stopped, the sun came out, my spot at the festival went OK apart from when I just ever so slightly fell off the stage and everyone nodded, smiled and raised eyebrows politely at my garbled, babbly story of being on a bus that was struck by lightning. It's a story that's more Reg Varney than Roy Sullivan and it cured little more than a bit of rain-induced grumpiness, but by Jiminy, I was on a bus that was struck by lightning.

5

'PLAIN, USEFUL AND UNPRETENDING': THE UMBRELLA

There isn't much we can do to protect ourselves from lightning beyond staying indoors, wearing rubber boots and not ringing church bells. I was sure that, given that it is a comparatively easy thing to do, protecting ourselves from the rain had been a long tradition, which is why I was surprised to learn that we came late to the umbrella. You would think that, given the amount of complaining we do about the rain, not to mention the number of umbrellas in circulation as we plough further into the third millennium, we'd have been at the cutting edge of umbrella technology from the start. But no, we were relatively late adopters, and that we took to the umbrella at all is due largely to the persistence of one single-minded Georgian.

Even considering their late introduction to our society, umbrellas don't get nearly the credit they deserve. They perform an entirely selfless function and how do we treat them? We leave them on buses, on trains, in pubs, to the extent that an estimated eighty thousand umbrellas are left each year on London's transport system alone, almost half the entire annual total of the capital's transport lost property items. That's more than two hundred discarded umbrellas a day doing endless turns around the Circle Line or ending up homeless and unloved in some floodlit bus depot in the early hours of the morning.

'Plain, useful and unpretending, if any of men's inventions ever deserved sincere regard the umbrella is, we maintain, that invention,' wrote William Sangster in his 1855 book, *Umbrellas and Their History*. And he should know: not only was he enough of an authority to be the author of what is currently the only book devoted to the brolly story, he was also among the leading umbrella-makers of nineteenth-century London (and the first to utilise the waterproof properties of alpaca wool, no less). Sangster is undoubtedly right, even if he does spend most of his book finding ever more long-winded ways of making exactly this point. The umbrella is a brilliant piece of technology, one that has an important everyday use and one that cannot really be improved upon. When in action the umbrella spreads its 360-degree wing to protect the user from the rain, and when not required it can in many cases be reduced to the size of a banana and dropped into a bag until it's needed again. You don't need to plug in an umbrella. It never needs recharging. It can be as cheap or as expensive as you like. You can't download an app to replace the umbrella.

Robert Louis Stevenson shared Sangster's admiration for the brolly. 'It is the habitual carriage of the umbrella that is the stamp of respectability,' he said. 'The umbrella has become the acknowledged index of social position.'

Stevenson backed up his claim by citing the example of Daniel Defoe's most famous creation, Robinson Crusoe, who was given one of the first great literary umbrellas, albeit one fashioned from branches and leaves.

'Crusoe was rather a moralist than a pietist,' said Stevenson, 'and his leaf umbrella is as fine an example of the civilised mind striving to express itself under adverse circumstances as we have ever met with.'

Indeed, the commemorative plaque in Hull – from where Crusoe originally sets sail in the novel – shows him with his musket in one hand and an umbrella in the other; a balance between civility and survival. We barely notice his umbrella in the book today, but in the eighteenth century 'robinson' was a common term for the umbrella.

The earliest references to umbrellas go way back to ancient Persia and Nineveh. These first manifestations were designed to keep off the searing heat of the sun rather than rain, and then they're only seen being held over the heads of monarchs and dignitaries in ancient illustrations. In the fifth century BC there are accounts of Greek women carrying parasols and the same later on, in the Roman Empire. In India and China at around this time, the parasol approached the form we know in today's umbrellas, and appeared to be collapsible judging from the illuminations and drawings that survive. As it was in the civilisations further west, the parasol seemed to be a symbol of power and status.

Both parasol and umbrella seemed slow to seep further across Europe, however. Thomas Coryat, in one of the earliest travel books ever written in Europe, his *Crudities* of 1611, noted of Italians that:

Many of them doe carry other fine things of a far greater
price, that will cost at the least a duckat, which they commonly call in the Italian tongue umbrellas, that is, things
which minister shadowe to them for shelter against the
scorching heate of the sunne. These are made of leather,
something answerable to the forme of a little cannopy, &
hooped in the inside with divers little wooden hoopes that
extend the umbrella in a pretty large compasse. They are
used especially by horsemen, who carry them in their
hands when they ride, fastening the end of the handle
upon one of their thighs, and they impart so large a
shadow unto them, that it keepeth the heate of the sunne
from the upper parts of their bodies.

This is the first use of the word 'umbrella' in an English text,
so we can thank Coryat for its introduction (we can also thank
him for the table fork, which he also introduced to the nation
after acquiring one on his travels). The word itself is derived
from the Latin *umbella*, meaning 'a little shadow', and the Italian
word *ombra*, meaning 'shade'.

By the early eighteenth century the umbrella had become
associated more with the rain than the sun: a 1708 dictionary
defines it as a 'screen, commonly used by women to keep off
rain'. As the century progressed umbrellas were to be found
almost exclusively by the doors of coffee houses, inns or dining
rooms as staff would use the umbrella to shelter departing clients
from the rain between the premises and their waiting carriages.
If a man was required to walk in the rain then he would simply
wrap his cloak around him to keep out as much of it as possible: umbrellas were for servants and footmen to hold over the

heads of the great and good for short distances, not for general use.

The umbrella needed a talisman; it needed someone to grasp it by the handle and unfurl it over society, to waggle its dripping canvas over the head of social convention and poke snobbery in the eye with its tip. That talisman arrived in the second half of the eighteenth century, in the eccentric form of Jonas Hanway.

Hanway deserves to be remembered for many things (a noted philanthropist, he was behind a number of incredibly good works) and not so much for others (he opposed the naturalisation of Jews). It would probably send a furrow of bemusement across his brow if he knew he'd been remembered best for defying social norms simply by using an umbrella. It's as if people would now say of Dr Barnardo, 'Ah yes, the man who wore galoshes,' or William Wilberforce being remembered for a prototype cagoule. But for all his quirks, faults and eccentricities, Jonas Hanway is definitely one of our weather heroes.

He was born in 1712 in Portsmouth, where his father was in charge of the naval stores until he was killed in an accident when Jonas was still a small boy. The family moved to London to stay with relatives, where Jonas was introduced to the ways of business from an early age. At seventeen he was apprenticed to a merchant in Lisbon and in 1743 he was appointed agent for Persia by the Russia Company of St Petersburg, the new capital of Imperial Russia that had been built from nothing by Peter the Great on the banks of the Neva barely forty years earlier. No sooner had he arrived in the young city than Hanway set off on a perilous journey to Persia with twenty carriages of English linen, a journey that would prove to be the making of him.

On arriving in Astrabad he was detained after being caught up in a rebellion. Eventually he escaped, minus his cargo, and was struck down with snow blindness, recovering just in time to see his tent burn down around him when a musket went off by accident and set fire to the roof. Somehow he still managed to regain most of his lost cargo, deliver it safely and return to St Petersburg. After spending a further five years on the banks of the Neva he headed back to London in the autumn of 1750 and moved in with his sister. You'd have thought that having endured such adventures Hanway would have been a much sought-after storyteller at a time when most foreign lands were unknown to London society. Indeed, the bandit-defying traveller often bemoaned the lack of interesting company in London and as a result felt that he had to furnish his dining room with curious objects, writing:

> I found that my countrymen and women were not as fast with the art of conversation; I have therefore presented them with objects the most attractive that I could imagine and such as cannot be easily imagined without eliciting amusing, exciting discourse; and when that fails there are the cards.

Unfortunately, if his written account of his adventures is anything to go by, Hanway was an extraordinarily tedious old windbag. Having read the book he produced of his travels, take it from me: the man could reduce the most rip-roaring tale of derring-do to the excitement level of a corporate accountancy textbook. The cards, I imagine, were rarely long in being produced.

But I'm perhaps being a little harsh. When published, Hanway's travel tales sold well enough for him to become well known in Georgian society and he remained successful in business, setting out each day to the St John's coffee house near the Royal Exchange to conduct affairs on behalf of the Russia Company.

Which is where the umbrella most likely entered the story. Hanway is credited as the first person in London to carry his own umbrella, something that doesn't sound all that revolutionary or inspiring but which at the time would have caused a sensation in a society that functioned on appearance and etiquette. As we have seen, the umbrella was the preserve of servants to keep their betters dry on their way to their carriages. For a gentleman, not least one as notable as Jonas Hanway, to be seen carrying one in the street would have set lips pursing, eyebrows rising and tongues wagging. It just wasn't done. It made people think you couldn't afford a carriage. The carriage-men looked askance as they feared a loss of livelihood if other methods of shelter became popular, and they did not hold back in their condemnation. A footman named John MacDonald recalled in a 1770 memoir that when he ventured out with an umbrella made from Spanish silk, and hence an obviously foreign import, the coachmen would call out, 'Frenchman! Frenchman! Why don't you call a coach?' The social minefield through which Hanway was marching is also described by the clergyman John Newton (who wrote 'Amazing Grace', among other things), who said at around the same time, 'To carry an umbrella without any headgear places a fellow in a social no man's land in the category of one hurrying round to the corner shop for a bottle of stout on a rainy day at the behest of a nagging landlady.'

Hanway could have been committing social suicide but clearly didn't care (the fact that he also still carried a sword, which had long gone out of fashion, may also have reduced the chances of direct opprobrium). As we know, he considered most of London society terrible bores anyway, but most of all he had other more important things on his mind than how he looked when beetling to the coffee house in the rain. After all, most of the time he was setting out to do some of the good works for which he should chiefly be remembered.

In 1756 he helped to set up the Marine Society in order to help provide men for the Royal Navy without resorting to the press gang. Britain was an enthusiastic participant in most of the naval wars during the eighteenth century and a good supply of sailors was becoming ever more important. Who knows, maybe Hanway carried harrowing memories from his childhood in Portsmouth's naval dockyard of men forced into service, miserable wretches lined up on the dockside destined for a life at sea in which they'd had no say. Maybe he wanted to do something in his seafaring father's memory (he would also become victualling governor to the Navy, a clear echo of his father's post in Portsmouth). Either way, through the Marine Society Hanway helped to ensure that young men joining the Navy were given clothing, food and education; that they were encouraged to join via better conditions rather than forced into service through falling into conversation with the wrong bunch of lads after one ale too many.

Hanway became a governor of the Foundling Hospital, which had been set up in 1741 by Thomas Coram to provide a home for deserted children, and was also instrumental in setting up the Magdalen Hospital for the Reception of Penitent

Prostitutes in Prescot Street, Whitechapel. Even at the end of his life, Hanway was campaigning on behalf of child chimney sweeps.

The progress of Hanway's good works was hampered, to his mind, by the effects upon society of the widespread drinking of tea, something for which he had an almost pathological hatred. He blamed the substance for most of the ills afflicting society – and this in the heyday of the gin shop. He published a series of letters on the subject at the end of a typically impenetrable account of a journey from London to Portsmouth. The letters were addressed to two unnamed women, and no replies are published: there's every chance that the letters were greeted with a drooping of the shoulders, a roll of the eyes and quite possibly a 'Why can't he just leave me alone?'

Tea vexed Hanway greatly. 'To what can we ascribe the numerous complaints which prevail?' he asked with uncharacteristic brevity. 'How many sweet creatures of your sex languish with a weak digestion, low spirits, lassitudes, melancholy and twenty disorders which in spite of the faculty have yet no names except the general one of nervous complaints? Let them leave off drinking tea.'

It wasn't just women: 'Men seem to have lost their stature and comeliness; and women their beauty. I am not young but I think there is not quite so much beauty in this land as there was. Your very chambermaids have lost their bloom, I suppose by sipping tea. What is it? An idle custom, an absurd expense tending to create fantastic desires and bad habits in which nature has no part and which render us less happy or more miserable than we should otherwise be.'

Aversion to tea notwithstanding, Hanway persisted with his umbrella for thirty years, becoming a familiar sight on the streets of London. Once he'd broken the taboo others took to the umbrella, and by the time Hanway died in September 1786 they were a common sight on the streets of the capital. He was a true pioneer, albeit an unwitting and unintentional one. He wasn't doing anything particularly brave, just staying dry as best he could. He probably barely even noticed the attention, so pre-occupied would he have been with thoughts of business, charity and the evils of lapsang souchong.

Although he's buried in Hanwell, west London, Hanway's charitable works earned him a posthumous plaque in West-minster Abbey. His memorial is in the western aisle of the north transept and portrays Britannia handing clothing to boys embarking on a career in the Navy. Few tourists give it a second look but I paused in front of it and noted the inscription, which proclaims how his 'name liveth and will ever live whilst active piety shall distinguish the Christian: integrity and truth shall rec-ommend the British merchant: and universal kindness shall characterize the citizen of the world'.

There's no mention of his umbrella.

It took a step forward in design in the middle of the nineteenth century to really establish the umbrella as a true protector of the people. Until then umbrella frames were usually made of wood, making them difficult and impractical to mass-produce and thus ensuring they remained within the realm of the wealthy, but in 1851 the introduction of the Paragon frame by Samuel Fox revolutionised brolly production. Fox's steelworks at Stocksbridge, near Sheffield, was home to Fox Umbrella Frames

Ltd, from where he produced an umbrella of unprecedented design that could be manufactured on a scale that made its price accessible to more than just the privileged. Its u-shaped string steel structure was a beautiful piece of engineering and its basic mechanics, responsible for the graceful spreading of the canvas as the runner slides up the tube, are still in use in every umbrella today.

During that same year a sumptuously decorated umbrella belonging to the Maharajah of Najpur was displayed at the Great Exhibition at Crystal Palace as a symbol of wealth, importance and power. Many of those filing past it may well have had a brand-new Fox's Paragon tucked under their arm or sitting waiting for them in the cloakroom. Never would the history of the umbrella in these islands have a more poignant illustration of the shift in its standing. Behind the rope, a work of art, the umbrella as status symbol. In front of the rope, people carrying a symbol of Victorian design, engineering and changing social mores, a mechanical marvel of moving parts that required neither coal nor gas to run, just one hand to steady and one to operate, that worked with grace and beauty to provide its user with complete protection from the rain.

Whether Jonas Hanway would have approved of the Maharajah's Great Exhibition brolly, who can say. His travels had given him a taste for curious objects in the hope of prompting the fascinating conversations that he craved, while at the same time Najpur may have been a little too close to tea production for his liking. He'd certainly have looked kindly upon the Paragon, however, and may have raised a wry smile at how the new railways that delivered many of the Paragon-toting visitors to the Great Exhibition had finally put his old nemesis, the sharp-tongued coachman, out of business for good.

Before long the umbrella had become so common that tales of other uses than its intended one began to crop up. Sangster, writing at the very start of the umbrella boom, cites this tale of British stiff-upper-lippery from the heart of the Empire:

> Members of a comfortable picnic party were cosily assembled in some part of India when an unbidden and most unwelcome guest made his appearance in the shape of a huge Bengal tiger. Most persons would, naturally, have sought safety in flight and not stayed to hob-and-nob with this denizen of the jungle; not so, however, thought a lady of the party, who, inspired by her innate courage, or the fear of losing her dinner, perhaps by both combined, seized her Umbrella, and opened it suddenly in the face of the tiger as he stood wistfully gazing upon brown curry and foaming Allsop. The astonished brute turned tail and fled and the lady saved her dinner.

As well as frightening off peckish big cats with a simple unfurling, the umbrella is one of the few offensive weapons that one can carry with impunity. Indeed, *Broad-sword and Single-stick*, a self-defence manual published in the late nineteenth century, contains, in addition to 'Chapters on Quarter-Staff, Bayonet, Cudgel, Shillalah, Walking-Stick, and Other Weapons of Self-Defence', a whole section on the umbrella. Authored by Rowland George Allanson Allanson-Winn, along with Clive Phillipps-Wolley of the Inns of Court School of Arms and author of the two-volume *Big Game Shooting*, the manual is a spectacularly entertaining publication whose brisk tone makes it all the more enjoyable and which makes one consider the brolly, or gamp (after Mrs Gamp, the umbrella-toting midwife from *Martin Chuzzlewit*) in a way it has rarely been considered before:

> As a weapon of modern warfare this implement has not been given a fair place. It has, indeed, too often been spoken of with contempt and disdain, but there is no doubt that, even in the hands of a strong and angry old woman, a gamp of solid proportions may be the cause of much damage to an adversary. Has not an umbrella, opened suddenly and with a good flourish, stopped the deadly onslaught of the infuriated bull, and caused the monarch of the fields to turn tail? Has it not, when similarly brought into action, been the means of stopping a runaway horse, whose mad career might otherwise have caused many broken legs and arms?

To which the only response can be a breathless 'Why, yes! Yes it has!'

The manual advises two ways of deploying the umbrella when confronted by an assailant: using it in the manner of a fencing foil, and 'grasping it firmly with both hands, as one grasps the military rifle when at bayonet-exercise'. The assaulted umbrella-carrier is advised to not 'be the least squeamish about hurting those who will not hesitate to make a football of your devoted head should it unfortunately be laid low,' but there's also a cautionary tale:

It is, of course, an extremely risky operation prodding a fellow-creature in the eye with the point of an umbrella; and I once knew a man who, being attacked by many roughs, and in danger of losing his life through their brutality, in a despairing effort made a desperate thrust at the face of one of his assailants. The point entered the eye and the brain, and the man fell stone dead at his feet. I would therefore only advocate the thrusting when extreme danger threatens – as a dernier resort, in fact, and when it is a case of who shall be killed, you or your assailant.

Probably the most famous incident of umbrella-related death occurred in London in 1978, but it involved neither pierced brains nor uncouth kickings in back streets. Georgi Markov was a dissident Bulgarian novelist and playwright who had defected to the West in 1969, while visiting his brother in Italy. He eventually settled in London, becoming a journalist with the BBC World Service and also broadcasting on Deutsche Welle and Radio Free Europe, where he was constantly critical of the Bulgarian government, and its leader Todor Zhivkov in particular. On 7 September 1978 – Zhivkov's birthday – while on the south side of Waterloo Bridge, waiting for a bus to take him to his shift at the BBC, Markov suddenly felt a sharp pain in the back of his right thigh and, on turning round, saw a man directly behind him pick up a dropped umbrella, apologise, run across the road, jump into a taxi and disappear into the London traffic.

The next day Markov went to hospital with a fever and within four days he was dead from ricin poisoning. A tiny pellet no bigger than a pinhead, in which two holes had been drilled, was found inside a puncture wound in his thigh. Experts at Porton Down suggested the pellet had been filled with ricin and a glucose substance that disintegrated at body temperature had been smeared over the holes. Once the pellet had been fired into Markov's leg and the ricin having made its way into his bloodstream, there was nothing that could have saved him.

No one has ever been convicted of the murder, but it seems certain the weapon was the umbrella that Markov had seen the man pick up. The Bulgarian secret police, the Darzhavna Sigurnost, carried out the hit, for which an Italian-born Danish citizen known as 'Agent Piccadilly' remains the prime suspect. General Vladimir Todorov, the former chief of intelligence, was sent to prison in 1992 for destroying ten volumes of material related to Markov's assassination. Later the former deputy interior minister Stoyan Savov killed himself when about to stand trial for his part in covering up the assassination. Most tantalisingly of all, when the Bulgarian government fell in 1989 a stock of umbrellas modified to fire tiny darts and pellets was found in the interior ministry building. It's a story that has terrific ingredients: the Cold War, the dissident, the assassin, Waterloo Bridge, an escape in a black taxi and the umbrella, that most mundane of objects, as the murder weapon.

There have been other famous umbrellas. Mary Poppins's brolly with its talking-parrot handle was, when unfurled, an instrument of magic: there are few greater images from childhood than Julie Andrews, suspended from her umbrella, descending gently onto the streets of London. When her brolly was tightly furled, however, there was a strictness about the nanny, a primness that made it instantly clear there was to be no funny business.

At a rainswept Wembley Stadium in 2007 the England football manager Steve McClaren assured his own demise when, while watching his team lollop to a gloomy defeat to Croatia that meant England failed to qualify for the European Championships, he stood dwarfed by an enormous golf umbrella, impotently surveying the carnage in front of him. It was a terrible error of judgement, but then so was most of his tenure in charge of the English national side. For one thing it disunited him from his team – if they were expected to do their jobs unprotected from the rain, then why should he be any different? For another, it was a barrier both physical and psychological between him and the crowd. It was if he was acknowledging they were on his back, and by holding the umbrella it was as if he was hiding from them.

Who knows, if he'd taken the umbrella into the technical area but kept it closed things might have been different. It would have said to the players and supporters, I'm in this with you. Look, I've got an umbrella that could shelter me from all this but I'm not going to use it because we're all in this together. It would have sent out a message that combined solidarity with authority – there is something about a person brandishing a furled umbrella that gives them an air of trustworthiness. It suggests they're equipped for adversity. It suggests preparation, organisation, a readiness for all outcomes. The image of Steve McClaren standing alone at Wembley with his golf umbrella suggested exactly the opposite. It managed to look both clownish and arrogant at the same time. The umbrella was a parachute: he'd bailed out. It was probably the most obvious psychological message in the history of sport, the umbrella as white flag. That night in north-west London Steve McClaren and his umbrella were about as far from Robinson Crusoe's defiant expression of rationality and dignity as you could possibly imagine.

Stroll down any major thoroughfare after a storm and I guarantee you'll see one recurring image: the broken, spindly remnants of an umbrella sticking out of a bin. The worse the storm, the more you'll pass. When you see them like that, their skeletons fractured and broken into ugly, pathetic shapes, fabric torn and flapping, you realise just how graceful and elegant the umbrella really is, even the cheapest one; even the one with a garish pattern all over it, or the logo of some smug corporation on every other segment. There's a wonderful dignity about the umbrella, with its smooth, symmetrical flowering as you put it up, the effortless movement and coordination of countless working parts, the elegance of its dome – the umbrella is a beautiful machine. Seeing one battered and ruined and shoved unceremoniously into a bin should sink the heart of anyone concerned with our relationship with the weather.

From expertly crafted bespoke affairs of the finest imported woods that cost hundreds to the tiny cloudburst-induced buy from a basket inside the shop door with a price in single figures markered onto a fluorescent star, the umbrella is a triumph of nearly every kind of human industry. Its design is simple but beautiful, as timeless and changeless as every perfect thing; its physics and construction are equally simple and, crucially, easy to operate. Its purpose, this canopy of benevolent beauty, is to protect us from the worst of the weather, a task it performs uncomplainingly, to order and with a minimum of fuss.

For all its flawless design and operation, however, we will be leaving our umbrellas on bus seats, under pub tables and in overhead train racks for as long as the rain keeps falling.

6

THE NUMB EXTREMITIES OF RENÉ DESCARTES: HOW WE STARTED THINKING ABOUT THE WEATHER

This was the worst part: getting out of the carriage. Being roused from his bed at four o'clock in the morning just once would have been torture enough for the old thinker, but this was happening every day, in the darkest, most frozen months of winter. He'd made his reputation from having some pretty startlingly good ideas, but moving to Stockholm definitely wasn't one of them. The coachman wrenched open the door and the familiar feeling of the icy air probing his cloak for a way in and the snowflakes alighting on his face made him suck in his breath. René Descartes stood up, wrapped his cloak around himself as tightly as he could and stepped down into the snow. The force of the freezing wind caught him full frontal and he trudged, gasping, away from the coach, crossed the bridge and hammered on the door.

He stood in the snow-muffled silence unable to feel his fingers or toes and breathing in short gasps that sent small clouds up into the night sky. The door opened and the bleary face of the guard at least confirmed that Descartes wasn't the only one to regard this hour as the middle of the night. Recognising the philosopher, the soldier nodded and ushered him inside, closing the door quickly and shooting the bolts. Descartes stamped the snow from his boots and brushed it from his cloak. Everything was numb; even his mind felt as if it were being throttled by icy fingers.

'I think that in winter here men's thoughts freeze like water,' he said. The guard just shrugged: it was one of the Frenchman's favourite lines and he used it often. Reluctantly Descartes took off his cloak and hat and handed them to the guard before commencing the familiar journey up the staircases and along the corridors until he arrived at the usual place. The guards stood aside from the big wooden door with its studded iron hinges and he knocked three times.

'Come in, Monsieur Descartes.'

He paused, took a deep breath, lifted the latch and entered the library. There was the queen, as bright-eyed as ever, sitting at the table with the books open and her hands in her lap, looking at him expectantly. The queen was hardy to the point of inhumanity: the room was, as usual, freezing save for the puny warmth sputtering from the candles on the table and a few small token flames in the fireplace on the far side of the room. Christina slept no more than five hours a night and saw fit to schedule their classes for five o'clock in the morning, the antithesis of Descartes's favoured working method of remaining in his bed until noon, thinking and writing.

He wished he'd never come to Sweden. With good reason, too: he didn't know it then, but these early starts really were killing him. He had only a few weeks left to live. The weather would ultimately be responsible for his demise; the same weather he had sought to explain in terms that would finally shake off the shackles of the ancient scholars, the same weather whose temperatures and air-pressure readings he had been faithfully recording since arriving in Stockholm.

'So,' said the queen, 'what are we going to talk about today, Monsieur Descartes?'

René Descartes was one of the greatest thinkers of all time but he made some shockingly bad decisions. On the face of it, accepting a job tutoring the Queen of Sweden in philosophy looked like the perfect posting. Christina was a highly intelligent woman fuelled by strong intellectual curiosity, who had set about turning her court into a renowned centre of scholarship, surrounding herself with some of Europe's finest poets, musicians, writers and philosophers. Hence her invitation to Descartes and his subsequent setting-out on the long land and sea journey to Sweden in September 1649 in good spirits.

He couldn't have known, but even as he set out by sea from Holland things were already going badly wrong. Between issuing the invitation and Descartes's arrival, Christina's enthusiasm for Cartesian philosophy had waned. She had discovered the delights of Ancient Greece and was setting about pursuing her studies in a more Hellenic direction. This would be the worst possible news for Descartes: he'd long regarded the classics as useless and had spent much of his career dismantling as much of the wisdom people had derived from the Ancient Greeks as possible.

Nonetheless, the queen was still excited by the philosopher's arrival, sending for him the morning after he'd made landfall in Sweden and offering him a six-week tour of the country, so that he might familiarise himself with his new home. Having just endured a long and uncomfortable journey from the Low Countries, Descartes politely declined. Christina's next flash of inspiration was that he should take part in the new ballet she'd commissioned to celebrate the Peace of Westphalia, which had a year earlier ended the Europe-wide devastation of the Thirty Years' War.

Now, those of us who were forced to spend a proportion of our late teens studying Descartes's writings at university might have relished the thought of the obsequious old sourpuss capering about in a pair of tights in front of a sniggering audience drunk on peace and akvavit – I, for one, would have attended carrying a paper bag full of gobstoppers with the sole intention of pinging at least one off his codpiece with a well-aimed fizzer from the dress circle – but alas Descartes managed to talk the queen out of his actual performing, reducing his participation to writing the libretto. Despite the ballet's roaring success, Descartes realised his move to Sweden really was turning into a nightmare.

Relief of a sort came when Queen Christina decided that she actually did want philosophy lessons after all, but Descartes's jaw would have clunked open when he learned that these were to take place at five o'clock in the morning during Sweden's coldest and darkest months. While worried about the climate before coming to Sweden, he had surmised that if anyone would know how to keep warm in temperatures well below zero it would be the Swedes. However, he'd been quite unprepared for just how cold it was in Stockholm.

The queen seemed somehow oblivious to both the cold and the uncivilised hour: for Descartes, the best thinking was done in warm, comfortable surroundings (indeed, he'd come up with his famous *cogito ergo sum* while shut away in a tiny room furnished with only a heating stove). It took only a couple of weeks of this horrifying nocturnal routine for the effects to take their toll on Descartes. He fell ill but, as a Catholic in a determinedly Lutheran country not long after a devastating religious war that had lasted for a generation, Descartes didn't trust the Swedish physicians and resorted to a dubious-sounding medication of his own devising: liquid tobacco taken in warm wine. Despite his faith in this remedy, by the end of January 1650 his health had worsened and he was suffering from pneumonia. He attended a church service at Candlemas on 1 February and it was clear the man was very ill indeed; he finally allowed the queen's physician to treat him.

After a week he seemed to have recovered a little – enough even to talk philosophy with a friend for a while – but when he decided he was well enough to get out of bed and sit in a chair he collapsed into the arms of his servant. 'This blow means I'm leaving for good,' he said when he came round, and at four o'clock on the morning of 11 February 1650 he died. Descartes's climate-related demise was poignant given that he was arguably the most famous of a string of thinkers who set themselves the task of explaining what appeared unexplainable: the weather. Why did it rain? Why didn't it rain? Where does wind come from? Were thunder and lightning truly the wrath of God?

Descartes had put his considerable mind to this in *Les Météores*. He made a decent fist of it too: while his theory that thunder was unconnected to lightning and came about as a result of the clouds bumping into each other was pretty wide of the mark, his work in optics and refraction meant that he had a pretty good stab at explaining rainbows. But what really made Descartes so significant to the history of the weather were the detailed records he kept during his time in Sweden. He was an early adopter of the barometer and the thermometer, both of which were becoming popular in scientific and academic circles, meaning he could keep reasonably accurate measurements of temperature and air pressure. He'd noticed in 1648 that levels of mercury in his Torricelli tube – the earliest form of the modern barometer – rose and fell in line with changes in the weather. Descartes joined with the French polymath Marin Mersenne, whom he had known for many years, and the young Blaise Pascal, who had been introduced to him by Mersenne as they had a mutual fascination with Evangelista Torricelli's experiments in air pressure and recording the weather. They formed a trio of weather-recording boffins marking off their mercury readings on special pieces of paper two and a half feet long that Descartes had devised for the task: in a sense, the first weather charts.

Even though many of his weather theories turned out to be incorrect, Descartes more than earns his place among the pioneers of the science of meteorology simply by turning his attention to it, seeking to dispense with the philosophical shackles that had been attached for centuries and lead them in a more scientific direction. Descartes was one of the first to sit down and truly *think* about the weather rather than merely observe it. This is why Descartes forms a crucial part of the legacy to which we are indebted for our understanding of what goes on in the air around us. It's a history that features some of the greatest names in philosophy and science, most of whom you would never have associated with meteorology.

Even the man who arguably started it all is someone you'd never think of as, say, the Michael Fish of his day. He wouldn't even have thought of himself as a meteorologist, despite being directly responsible for the term. The book containing his theories can be hard to find, being one of the most minor of even his minor works, yet Aristotle instigated some of the most remarkable advances in the philosophy of weather, taking us away from mythology towards more rational explanations. He may not have been right about most of them, but he was convincing enough to rule the weather roost for the best part of two millennia, giving us a basis from which to work on the philosophy and science of weather, and ensuring that we didn't have to keep swallowing legends like that of Phaeton, for example.

When I was growing up my dad worked for an airline in an office in London. He'd joined as a tea boy when he left school and had worked his way up to quite a senior position, doing something that, at the age of seven, I didn't remotely understand. However, the fact that I didn't know what he did still didn't excuse how I managed to convince everyone at school that he was a Concorde pilot. From staff room to playground the whole place was abuzz with the fact that my dad flew Concorde, which was then in its early days of service and causing awed faces to swivel skyward whenever it passed overhead, all sleek glamour and effortless cool. Concorde used to fly over our part of London at around ten past six every evening, and if I happened to be having my tea at a friend's house I'd pause, cock an ear to the distinctive sound, give a wistful sigh, say, 'There goes Dad', and fork another potato croquette into my mouth.

I'm not sure how long I managed to keep up the pretence — it seemed like ages; in reality it was probably no more than a few weeks – but my already brittle story was finally shattered at parents' evening when my teacher nervously asked Dad if he'd consider coming in and giving a talk to the whole school about being a Concorde pilot. If the hefty lenses in his glasses hadn't already raised doubts that I might be full of shit, Dad's reply, along the lines of 'What on earth would I know about being a Concorde pilot?', sealed my fate. The game was well and truly up and my standing among my peers took a tumble from which it has never since recovered.

It was playground bullshitters like me that made it difficult for kids like Phaeton, kids who had a parental story so genuinely impressive that no one believed them. In Phaeton's case his dad was Helios, the Greek god of the sun and a permanent fixture in the premier league of gods ancient and modern. The trouble was that nobody believed Phaeton when he told them this, as his Oceanid mother Clymene was married to Merops, a mortal. No matter how often he assured them that his dad was the sun god, no matter how high-pitched his protestations became, the other kids weren't buying it. He'd go home, toy with his food and sulk until finally his mother asked him what was wrong. 'It's the kids at school,' he'd say, 'I tell them who Dad is and they won't believe me. How can I prove to them that Dad is who I say he is?'

His mother could probably appreciate the position in which her son found himself; there's a good chance her friends had reacted in the same way when a nervous Helios had first taken her to the pictures. Sympathising with the boy's plight, she recommended that he travel east to the palace of the sun god and explain the problem to his father. Helios was delighted at the unexpected visit of his offspring, but reacted in a manner that

can only be described as – heh – hot-headed. Rashly, he immediately granted Phaeton one wish, something he could facilitate
that would prove to the cool kids in the playground that his
dad really did outdo all the other dads. Phaeton's eyes lit up.
'What would be really cool,' he said, 'what would really
convince them I was the real deal, would be to let me drive the
chariot of the sun.' This was a big favour to ask, and not one that
Helios would be happy about granting but, being an important
god in the pantheon, he knew he couldn't really backtrack
on his reckless wish-granting without losing face. He was in
a difficult position: the route the sun chariot took across the sky
from east to west on its diurnal round was a notoriously treach-
erous one and the horses were famously headstrong. He didn't,
as a rule, allow anyone else to drive the chariot, but this was his
own son. And he'd sort of promised. He tried to persuade the
boy to think of something else.

'Beware, my son,' he warned, 'lest I be the giver of a fatal gift.
Recall your wish while you yet may: it is not honour but
destruction you seek and I beg you to choose more wisely.'

His advice fell on stony ground and, just ahead of the break
of dawn the next morning, the horses were harnessed to the sun
chariot and Phaeton climbed aboard. Helios advised Phaeton to
use the whip sparingly, to keep a tight grip on the reins and to
stick to the 'middle zone' of the sky, the route that kept the temperature and conditions below at most favourable levels.

As soon as they set off, however, the horses sensed that this
wasn't Helios driving. Indeed, it was somebody not nearly as
strong as the sun god; someone who they realised would not
be able to keep them in check. With a terrified Phaeton hanging on desperately they rampaged across the sky and veered

sharply out of the zodiac, causing their passenger to drop the reins altogether. The horses hurtled hither and thither to previously unknown parts of the heavens. They thundered across the highest stars in the firmament and galloped down nearer to the earth than they'd ever been. Beneath them the clouds began to billow smoke, the tops of mountains caught fire and entire harvests burst into flames. Cities burned to the ground, whole nations were reduced to ashes and even Atlas himself passed out from the heat. Phaeton looked down in horror at the apocalypse he'd caused, gasping air that scorched his lungs and crying out for help. It was left to Zeus to save the day, shooting the sun chariot out of the sky with a bolt of lightning, halting the devastation but killing poor Phaeton. The destruction he had caused was incalculable and irreversible.

This is one our earliest weather-related tales: a great example of humans' ability to combine both mythology and a rattling good yarn in an attempt to explain things we don't understand, like the path of the sun across the sky and why some days are hotter than others. Maybe the story was based on some ancient cataclysmic heatwave that really did ruin crops and result in widespread deaths, but in any case, the story of Phaeton – and, more important, the belief that a sun chariot passed across the heavens each day – was one of the earliest and most imaginative attempts to explain the weather.

Then Aristotle came along and applied a scientific mind to the subject for the first time.

There have been few bigger brains than Aristotle. He was a thinker, teacher and writer in just about every aspect of learning: some even credit Aristotle with being the last person in history who knew everything there was to know. He lived in the fourth century BC, studied under Plato, founded the Lyceum, mentored Ptolemy and Theophrastus and tutored Alexander the Great. He also wrote. About everything.

Among his writings that survive – scholars believe that as much as two-thirds of Aristotle's literary output may have been lost – is the *Meteorologica*. It's not one of his famous works: even the scholarly introduction to the edition I have confesses that it's a little-read volume thanks to the 'intrinsic lack of interest' in its contents. If that really is the case, I would put up a spirited argument for a re-evaluation. It's not a hefty tome – my copy contains both the original Greek and the English translation which, combined, still come to less than four hundred pages – and is quite possibly the most important piece of weather-related writing ever, certainly at least until the composition of the Weather Girls' 'It's Raining Men'. It was important enough to remain the standard weather text for the best part of two thousand years, until the likes of Descartes began their attempts to construct a more scientific approach, and its name lives on in the very word we use for the study of weather: 'meteorology'. It's a landmark in the philosophy and science of the everyday; a brilliant attempt to explain the causes and rhythms of the weather. Most of it has now been proved to be incorrect, but *Meteorologica* should be the starting point for any discussion of weather. We owe everything from the shipping forecast to debates about climate change to Aristotle.

The groundwork had already been laid. A century earlier Empedocles introduced the theory that everything is formed by the four elements: earth, air, fire and water, whose mixtures were determined by the powers of love and strife. Empedocles clearly had a high opinion of himself: he died when, in an attempt to prove he was immortal and thus would rise up as a god after death, he threw himself into the bubbling crater of Mount Etna. After him came Hippocrates, the Great Physician, who supposed that, just as diet is crucial to physical health, climate is vital to the formation of personality and lifestyle. But Aristotle was the first to sit down and think about what makes the weather, where it comes from and where it goes.

Important though it is, *Meteorologica* is not exactly a gripping read. You can take it from me, as one of probably very few people who have actually read it. It's probable that Aristotle had never envisaged it as a collection of works gathered between hard covers and translated in such a way as to remove any sense of pleasure. Many of Aristotle's writings were designed with his students in mind; some are essentially little more than lecture notes. Old academic translations from classical Greek are never going to be particularly vivacious and are unlikely ever to be adapted for the stage as musical comedy, but beyond the almost Saharan dryness of the prose are some of the most important ideas and arguments in the history of weather.

Meteorologica is not just concerned with what we know as the weather. As well as proffering explanations for phenomena, Aristotle also writes about shooting stars, the Milky Way, earthquakes and comets: for him they were all connected. As he says early in the work, 'its province is everything which happens naturally but with a regularity less than that of the primary element of material things and which takes place in the region which borders most nearly on the movement of the stars'.

The vagueness and verbiage of this definition gives an idea of the tone of the translation, but what Aristotle is essentially saying is that *Meteorologica* is about 'you know, other stuff'. Even the title is vague: the word 'meteor' applied to things that were high up, making *Meteorologica* nothing more specific than 'some thoughts about high things'.

The basis of Aristotle's theories was situated in a fifth element far above the earth, in which the heavenly bodies moved. Aristotle called this the 'ether' and believed that the movement of objects within it – the stars and comets – was regulated. Below the ether were the four elements of air, water, earth and fire, which obeyed their own regulations. Of these he deemed fire the lightest and therefore the highest element, as flames are drawn upwards towards the heavens. The next level down was air, heavier than fire as it rose in water as bubbles. Next came water, because it fell through the air as rain, with earth the lowest, beneath the air. He believed the elements were constantly in motion: indeed, it was the air rubbing against the ether at about the height of the moon that caused the celestial fires that heated the planet.

His most confident assertion was on the cause of wind. Aristotle believed that the winds were an exhalation from the earth and came in two forms. The first occurred when the sun shone on hot, dry land, causing the earth to give off an exhalation of hot air that rose into the fiery heavens. Similarly, the sun shining on the sea led to a cool, moist exhalation; the moisture prevented it from rising as far as the dry exhalation: 'The moist exhalation does not exist without the dry nor the dry without the moist, but we speak of them as dry or moist according as either quality predominates.' These moist exhalations would reach a certain height and then become clouds, with the moisture falling back to earth as rain. Mist – what he called 'unproductive cloud' – was the residue of this process. Winds blew horizontally because the body of air surrounding the earth follows the motion of the heavens around the stationary planet; hence we feel the exhalations as they move in conjunction with the movement of the heavens.

While to modern minds Aristotle's theories are generally wide of the mark, this was the first comprehensive examination of what causes weather. After him came his pupil Theophrastus, who was so close to Aristotle he was given the old sage's library and manuscripts after his death and was his appointed successor in running the Lyceum. Theophrastus took Aristotle's ideas a little further. He was more forthright in asserting that the great weather events happened independently of the gods. As we have heard, lightning was believed to be Zeus's spear, and anywhere it struck immediately became a sacred place dedicated to appeasing his displeasure. Theophrastus asked, if lightning bolts were Zeus's punishment for some misdeed, why did they most frequently strike out of harm's way, in the middle of nowhere? To his mind, thunder and lightning were caused by the collision and division of clouds. Theophrastus also produced the *Book of Signs*, essentially a compendium of weather lore, but this was more prophecy than philosophy. 'It is a sign of rain when a crow puts back its head on a rock which is washed by waves,' for example. Likewise, expect rain 'if a hawk perches on a tree, flies right into it and proceeds to search for lice'. 'It is a sign of fair weather if a goldcrest flies out from a hole or from a hedge or from its nest,' apparently, while a donkey shaking its ears is an indication that there is a storm on the way.

Theophrastus aside, the science and philosophy of weather go quiet for centuries after Aristotle. Most people were happy to believe that the weather was governed by their particular deities, whether that be a bad-tempered Norse god grumpily lobbing thunderbolts around, or an omnipotent formless being whose whimsy decides the abundance of harvests. Big storms and great weather events were interpreted as the wrath or otherwise (but usually wrath) of God or gods, and there was no serious attempt to interpret weather in any more scientific way for more than a thousand years.

The Renaissance saw the first seeds of a scientific approach sown with Leonardo da Vinci's invention of the hygrometer to measure the level of moisture in the air in the 1400s, but it's to poor, frozen Descartes we owe thanks for constructing a bridge between the philosophical and religious schools of weather thought, laying the groundwork for the weather forecasts we know today. He's not often credited as such, but it's worth stressing that René Descartes is one of the most important figures in the history of meteorology.

When she heard of his death, Queen Christina immediately announced grand plans for a marble tomb among Sweden's kings in the Riddarholmen Church. In the short term, while the arrangements were being made, Descartes, as a Catholic, was buried in a public cemetery among the unbaptised. Alas Christina never quite got around to reinterring the old philosopher whose end she had hastened with her insistence on dragging him from his bed in the middle of the night and Descartes remained in unconsecrated ground for nearly eighty years before he was spirited away to a grander resting place in France. During that time the science of weather would make advances of which the old thinker could only have dreamed. The winds of change were not just blowing, they were being analysed and explained.

7

WHEN THE WIND BLOWS

I really should have noticed the silence. I'd heaved myself out of bed, padded blearily to the stairs, thumped down them and pushed open the kitchen door. I was first up. This was unusual. I filled the kettle, plugged in the power lead and clicked the switch. The little orange light didn't come on. I squinted at it and tapped it. Nope, nothing. There was no rushing noise of rapidly heating water either. Dammit, the kettle was broken. I opened the fridge. The light didn't come on. It wasn't making its faint, and oddly comforting, metallic whirring either. I tried the lights. Nothing. A power cut. Great.

We'd not lived in the house long, so maybe that's why I didn't notice the silence. We had moved from a busy main road where there was a constant roar of traffic to a brand-new house on a leafy suburban estate. There was nothing like the volume of traffic, but there was usually something going on at this time of day: a car engine running while the owner closed the garage, kids heading off to school, the battery whine of a milk float, the whistle of the postman. There was none of that, but given how through the fog of morning bleariness the only thing I could see was the word 'tea' it didn't register with me at all: all that had really sunk in was that we had a power cut. I trudged up the stairs, crossed the landing, pushed open the door of my parents' bedroom and announced – in that indignant teenage voice that suggested it was all their fault – that there was 'a bloody power cut'. I'd expected to impart this news to supine sleeping figures but instead they were both sitting bolt upright in bed and, even more surprisingly, my sister was sitting between them. They looked ashen.

'Wha . . . what's happened?' I asked.

'Didn't you hear it?' answered my mother.

'Hear what?'

'Look out of the window,' advised my father.

I walked across to the window and peered out. The wooden fences that separated our postage-stamp garden from those of the houses behind were all lying flat on the ground, their posts broken at the bases. There were shards of glass all over the garden, which was also scattered with broken roof tiles. Not a plant pot was standing; they too had been reassigned to locations across the grass.

'Didn't you hear it?' repeated my mother.

'No, I didn't,' I replied. 'What's happened?' It was the morning of 16 October 1987 and I had slept right through the most fearsome storm in nearly three hundred years.

In mitigation I'd had football training the night before so was tired anyway. I was also a seventeen-year-old boy: sleep was my favourite hobby. I had been faintly aware of a distant creaking noise during the night – evidently the gale-force winds trying to tear the roof off the house – but it hadn't been nearly enough to rouse me from slumber. I was The Boy Who Slept through the Great Storm of '87. It's one of my greatest achievements.

The storm left a terrible mess in its wake. Eighteen people were killed: five in the port at Dover; two firemen when a falling tree fell on their tender in Dorset; and a man in Hastings was killed by a beach hut being blown along the seafront. In the same town a chimney crashed through three floors of a hotel, killing one man and leaving another hanging from a door handle in his pyjamas, calling for help. There were similar stories right across the south of the country; destruction on a scale not seen since the Second World War. Shanklin pier on the Isle of Wight was completely destroyed, a cross-Channel ferry ran aground at Folkestone and at Harwich a ship being used by the UK immigration service to house Tamil refugees from Sri Lanka slipped its moorings, pitched and tossed in the open seas for two hours and then ran aground. Three million houses were damaged, and more than a hundred thousand were left without power, some for up to two weeks as the electricity board tried to repair power lines. It was an epoch-making natural disaster.

The other three members of my family had sat up most of the night listening to the tiles crashing and glass smashing amid the roar and howl of the fiercest gales of our lifetimes, wondering if the roof was going to stay on. My dad gave us a lift to school, driving slowly and carefully through near-empty streets. Trees lay across roads, just some of the fifteen million uprooted that night, and there was debris everywhere. Eventually we reached the school, only to find more carnage. Nearly every tree was down: one large specimen had fallen on top of the school's minibus, completely flattening it. On its way down, the tree had sheared the stone window sills in its path clean off but not broken a single pane of glass. A fraught-looking deputy headmaster came puffing out of the door, walked over to my dad, motioned him to wind down the window and said the magic words, 'The school is closed today.' And home we went, reversing the slow, wide-eyed drive past what looked frighteningly like the apocalyptic post-nuclear scenarios I'd scared myself silly with as a child of the Cold War.

My memories of the Great Storm of '87 aren't exactly shattering. I was seventeen, I'd slept through it, I saw a squashed minibus and a few fallen trees, then got the day off school. For me it wasn't so much a once-in-a-lifetime weather experience as The Day The Kettle Didn't Work.

Two days earlier a depression had formed off the coast of Portugal and begun to make its way slowly and steadily north. A cold front off the southern coast of Ireland was given a massive boost in power by the arrival of warm air from Africa and cold air from the Arctic. Where these two air masses met a frontal system developed, with the warm air being pushed above the cold and causing the pressure to drop. Large quantities of water vapour condensed and formed clouds, which prompted a massive release of heat energy that turned into fearsome winds and caused the air pressure to drop further. The depression and its associated winds moved into the English Channel and in the early hours of 16 October made landfall somewhere around Bournemouth. From there it forced winds of more than a hundred miles an hour across southern England before heading out into the North Sea above the Humber estuary and veering further north before blowing itself out.

There were extraordinary swings of pressure: a barometer on

an oil rig off the coast of Norfolk recorded a drop of 6.8 millibars in just three hours during the night, while at Southampton pressure rose by eleven millibars in an hour. At Reading there was a barometric pressure reading of 958.5 millibars, the lowest pressure recording on land in Britain for over a century. The air temperature fluctuated too, falling six degrees in one hour on the south coast. Having occurred over southern England the storm was given much more news prominence than if it had happened in, say, the Highlands of Scotland, but it was certainly one of the most significant weather events of the twentieth century.

While the storm largely passed me by, one person for whom the Great Storm was a career-defining moment was Michael Fish. It almost pains me to mention his name in connection with the storm as the poor man has been so vilified in the following decades.

'Earlier on today, apparently,' he'd said on the afternoon weather bulletin as the depression was looming off the coast of France, 'a woman rang the BBC and said she'd heard there was a hurricane on the way. Well, if you're watching, don't worry, there isn't.'

In the tabloid breast-beating that followed ('WHY WEREN'T WE WARNED?') Fish became by far the biggest scapegoat. Jokes, eye-rolling and finger-pointing about the accuracy or otherwise of weather forecasts have been a staple ever since they first appeared in *The Times* in the 1860s, but the fact that Fish had not only failed to predict the hurricane but had actively asserted that there wouldn't be one made it open season on the BBC weather man, one of the least offensive figures on British television. The poor man must still be reminded of it, and his exhortation to the concerned lady not to worry is a standard of clip shows even today. On the late-night BBC2 forecast, even as the storm was starting to whip its way along the south coast, Bill Giles was sure that the strong winds would stay away, though there would be stiff breezes through the Channel. Only the shipping forecast gave any hint of the 'severe gales' to come until, in the early hours, the Met Office began to issue weather warnings to a sleeping nation.

In some ways I feel sorry for Michael Fish. Here was a man who had presumably been captivated by the weather at a young age, so much so that he chose to make its prediction a career. He would have studied hard to become a meteorologist, and had forged a long and successful career out of a lifelong passion: he was a man who loved his job. There's no doubt he felt he had every reason to believe that there wasn't a hurricane on its way: his charts and data would have told him just that. In fact, the very idea seemed impossible. So when word filtered through to him that a woman had phoned the BBC because she'd heard about a hurricane he salted it away to mention in the broadcast. It was out of the question, there was no way we were going to see a hurricane that night. He probably thought he was setting her mind at ease and that it gave the rest of the forecast a nice context as it was going to be a bit windy. But not hurricane-force winds. Definitely not hurricane force.

In the days – and indeed years – that followed, Michael Fish must have felt betrayed. Not so much by the press or indeed the rest of the meteorological community, who seemed disappointingly happy to let him carry the can (Fish and Giles were exonerated in a subsequent inquiry, while the Met Office was criticised for its inflexibility and for relying too much on computers for its forecasts), but by the weather itself. He'd given the best years of his life to the weather; I imagine that as a boy he'd been thrilled by the movements of fronts across the Atlantic, exhilarated by watching storm clouds gathering, had felt moved by the sheer silence of a heavy fog, taken satisfaction from correct forecasts that made people's days easier. Yet for all his love and hard work in its name, here was the weather hood-winking him, outmanoeuvring him and hanging him out to dry. It was a merciless betrayal.

Having said that, if he hadn't mentioned the woman phon-ing the BBC none of the ensuing opprobrium and ridicule would have happened. But who was she? Who was this myste-rious person who'd heard about the hurricane that eluded the predictions of the finest weather minds in Britain? Perhaps appropriately for a storm that gave suburbia its most traumatic night since the Blitz, it turned out that the woman responsible for the call was from Pinner, that staple of the sitcom and right in the heart of Betjeman's Metro-Land. She was a solicitor's clerk called Anita Hart, she'd been planning a weekend caravan break in the country and had asked her son Gaon, a keen mete-orologist who was studying the weather as part of his geography degree at Manchester University, what the weather might be like that weekend. Gaon consulted his charts and discovered a deepening depression making for the Channel. It didn't look good to him, he told his mother as much and she thought it best to see what the experts thought. She took the telephone direc-tory off the shelf, leafed through to the relevant page, picked up the receiver and inadvertently played her part in dooming Michael Fish to years of ridicule.

If 1987 proved anything it was how the wind is easily the

most powerful force in the weather armoury. Lightning might strike, but it strikes a small target. Rain may cause floods but they're generally restricted to a relatively small area. When the wind gets going it sweeps along for miles, crossing borders and seas, bending branches and flattening grass and crops. The wind is our most constant weather companion: rain, snow, hail, even sunshine can come and go but there's always wind. Whether it's a light summer breeze or a winter gale it's a rare occasion when there's not a breath of wind.

The wind is the backbone of the shipping forecast; it's granted more detail than any other weather condition and is almost the very reason why there is a shipping forecast. The wind is the sailor's enemy as much as his friend: rain, sleet, hail and fog pose their own problems, but nothing like when the wind gets angry and people on the sea are at their most vulnerable. The great storm of 1703 is the perfect illustration, as while 123 people died on land, between eight and ten thousand died at sea as a result of the same winds. In 1362 there was the *Grote Mandrenke*, the great drowning of men, when a southwesterly gale swooped across Ireland, England and the North Sea, causing the demise of the English coastal town of Ravenser Odd, which now lies under the sea with the obligatory legends attached to every submerged community about the sound of church bells tolling from somewhere beneath the waves. In Ireland it was known as the St Maury's Day Wind and, among other things, inspired what's possibly the worst weather-related verse ever committed to paper:

In that same year – 'twas on Saint Maury's Day –
A great wind did greatly gan the people all affray;

So dreadful was it then, and perilous,
And 'specially was the wind so boisterous,
That stone walls, steeples, houses, barns and trees
Were all blown down in diverse far countrees.

These 'far countrees' suffered one of the greatest disasters in European history: the wind caused such a sea surge that the North Sea inundated the low-lying Netherlands and northern Germany. As many as eighty thousand died in the *Grote Mandrenke*, with roughly half the population of the coastal areas between Jutland and what is now northern Germany believed to have drowned.

The sea defines us as an island people. We respect the sea, we fear it, we farm it, we protect ourselves from it as best we can but we know we'll never either conquer it or subjugate it. The wind, however, has the measure of the sea, whipping it up into surges and waves and floods like the *Grote Mandrenke*, or the North Sea surge of 1953 that caused death and destruction along the eastern coast of England and burst the man-made dykes and flood defences of the Netherlands.

The wind is invisible and everywhere, with discernible personalities: the bully, the vandal, the mischief-maker. It flits and sweeps around us like an excited attention-seeking child, playfully tugging at our clothes and ruffling our hair and making it difficult to read maps or newspapers by battering the pages. In Greenwich Park I once saw the wind whip off a man's hat, flip it over and plonk it on the head of his wife (which is still, incidentally, one of the greatest things I have ever witnessed).

The wind is a paradoxical thing, a creature of habit that's prone to acts of spontaneity. Generally speaking, the wind is our friend, but it's one that is also prone to violent rampages of temper. It is dangerous when riled, and liable to take off over great distances at a terrific pace, apparently on a whim. The constant circuits of the wind around the globe have defined trade routes, aided conquests and put civilisations in touch with other civilisations both for better and for worse. As well as making the sails of a child's beach windmill turn, the wind has been one of the great determinants of global history.

It is no wonder then that every strand of human civilisation in every part of the world has assigned names and personalities to the winds. To the Ancient Greeks the north wind was Boreas, an old man (his name survives in the northern lights, the aurora borealis) who, being a cold wind, was usually pictured in furs and carrying the shell with which he'd make his fearsome howling sound. The south wind was Notus, portrayed as sweaty and sticky from all the moisture he'd gathered on his way up from the south across the Mediterranean. Notus was also the god of mist and fog, meaning he was feared by seamen and apparently revered by thieves and pirates as he allowed them to go about their nefarious business undetected. Zephyrus, the gentle westerly mentioned in the opening of *The Canterbury Tales*, was given a more kindly persona while as for Eurus the east wind, well, no one seemed to pay him much attention, but he did bring warmth and rain.

The best depictions can be found on the Tower of the Winds in Athens. Constructed some time around 100 BC, the octagonal tower is a wonderful example of an early weathervane. It was – and indeed still is – a sundial and at one stage it also contained a water clock. The Tower of the Winds is believed to be the work of the Macedonian astronomer Andronicus of Cyrrhus and it demonstrates a brilliant understanding of the weather. Each of its eight sides is topped with a relief sculpture of the named wind in whose direction it faces: Boreas in the north, Kaikias facing north-east, and continuing all the way around with Eurus, Apeliotes, Notus, Livas, Zephyrus and Skiron. Originally the twelve-metre-high tower was topped with a giant iron representation of Triton, the messenger of the seas, complete with staff and trident (like Boreas, Triton also carried a shell; his was to summon up angry and giant seas). Triton would turn with the wind until his staff pointed in the direction from which the wind was blowing. Again the choice of the messenger of the seas demonstrates how closely the sea has been associated with the weather since classical times. In a way, the Tower of the Winds was a prototype shipping forecast: it wasn't forecasting the weather, but Triton was at least pointing out whence came the wind of the moment.

For something so beguiling, the basic global principle of the wind is fairly straightforward. He may be best known for the comet that bears his name, but it was Edmond Halley who first deduced the scientific basis of wind and contradicted Aristotle's theory of exhalations. In the 1680s the proliferation of European explorers bustling around the globe in doublet and hose meant that Halley had access to wind information from across the world. It wasn't comprehensive, but when cross-matched with readings from the new-fangled barometer that everyone was raving about he was able to get a strong feeling for the make-up of the global wind system. Hence it was Halley who first realised that winds are born at the equator. The heat and solar radiation from the sun warms the air there, which rises to the ceiling of the atmosphere and spreads out, heading towards the poles. Nearby air down on the surface of the earth senses the low pressure at the equator and rushes to fill it from the north and the south, while at the same time being pushed westward by the rotation of the earth (a phenomenon known as a Coriolis effect). These easterly winds are the trade winds: north-easterly in the northern hemisphere and south-easterly in the southern.

Meanwhile, the air that rose from the equator and set off for the poles starts to run out of steam at around thirty degrees latitude and sinks towards the earth, forming what we know as the Tropics of Cancer and Capricorn. The sinking air dries out, becomes warmer and divides as it falls, with some moving back towards the equator and the rest continuing towards the poles. This moving air comes up against and mixes with cooler air from the north to form fronts, the main factor in determining the variations in and vagaries of our everyday weather and the cause of the hurricane that deprived me of my tea that autumn morning in 1987. These fronts were one of the last pieces to be slotted into the jigsaw of our understanding of the weather, having been discovered by a group of scientists under Vilhelm Bjerknes at the Geophysical Institute of the University of Bergen in the 1920s.

To understand fronts, it is necessary to get to grips with the concept of high and low pressure. When there's a centre of high pressure at ground or sea level a great big lump of air is sitting there, weighing things down. As the air sinks downwards it makes wind by pushing outwards from the ground or sea, like when a fat man sits on a beanbag. In areas of low pressure the opposite is the case: there's more air around the area of low pressure than there is actually in it so the winds rush to the centre in order to fill the gap. The Coriolis effect ensures that the winds hot-footing it from a centre of high pressure turn clockwise around the centre in the northern hemisphere and anti-clockwise in the southern as they're drawn into the centre of low pressure. As the Norwegian meteorologists discovered, fronts are bands of reduced pressure that spread out from centres of low pressure; they are barriers between different levels of higher pressure that cause the variations in our weather.

It's a fairly simple process that creates wind, but one that has enough random factors to make the weather and the winds impossible to predict with total accuracy. As Robert FitzRoy, founder of what is now the Meteorological Office, explained in his practical manual, *The Weather Book* of 1863, 'air presses on everything within about forty miles of the world's surface, like a much lighter ocean at the bottom of which we live – not feeling its weight because our bodies are full of air, but feeling its currents: the winds'.

FitzRoy touches lightly on one of the most fascinating aspects of the wind: its influence on people. In this, we are drawn back to Aristotle. His theories about our atmosphere being formed of four elements tied in with the idea of our health being governed by the balance of the four humours: the combination of the elements in different layers and quantities produced weather in the same way that the humours defined health, temperament and character. When the elements – the fact we still refer to the weather as 'the elements' shows just how ingrained Aristotle's theories remain in our thinking about the weather – were roughly equal the weather was moderate and unchallenging, but when one element came to dominate the others things started to go wrong. Storms and the like would ensue until the other elements restored a sense of balance. Also, given that man's four humours corresponded with the four elements, it was inevitable that climate and weather would influence us.

Aristotle's thinking also echoed Hippocrates, who was convinced that the weather influenced humans. Herodotus, in his *Histories*, had already postulated that '[i]n keeping with the idiosyncratic climate which prevails there and the fact that their river behaves differently from any other river, almost all Egyptian customs and practices are the opposite of those of everywhere else', before Aristotle took up the same dubious opinion that people who lived in cooler, sharper mountainous regions were usually courageous and hardworking, while those who lived where hot winds blew across the plains were more likely to be short and stocky, with an innate antipathy to hard work. Hippocrates, in *On Airs, Waters and Places*, found European characters to be diverse because the continent's climate was so varied but, he theorised, other places had a more regular climate and hence the character of their people was easier to stereotype. For example, in Scythia it was always winter, and cold fostered barbarity: the Scythians, he'd heard, drank the blood and ate the flesh of the dead. As for Britain and Ireland, well, we were the last outposts of humanity, had the worst climates and were therefore the very archetype of barbarity. Strabo, in his *Geographica*, shared this view, noting that the inhabitants of Ireland 'are more savage than the Britons, since they are man-eaters as well as heavy eaters ... they count it an honourable thing, when their fathers die, to devour them, and openly to have intercourse, not only with the other women, but also with their mothers and sisters'.

Hippocrates believed that only lands like the one he just happened to live in, 'bare, waterless, rough, oppressed by winter storms and burnt by the sun', could produce 'hard, lean, vigilant, energetic, independent, articulate, well-braced and hairy men'. Men just like Hippocrates himself. Funny that.

Aristotle picked up the Hippocratic ball and thundered off towards the anthropological posts. In his *Politics* he labelled the people of the far north as brave but essentially stupid: they enjoyed their freedom but the prevailing cold left them too dim to organise themselves into doing anything useful or imposing their will on others. In Asia, the generally warm climate enhanced creativity and intelligence but the people lacked the bravery of their northern counterparts, which made them vulnerable to conquest and again unable to bend anyone else to their will. And who should be in between, enjoying the best of both worlds? Why, the Greeks of course: intelligent and brave, they were in the perfect climatic region to dominate the rest of the known world. Presumably this part of his teachings stuck fast in the mind of Aristotle's most notable pupil, Alexander the Great, who established one of the largest empires in the ancient world, extending the already expanded borders of his father Philip of Macedon all the way to the foothills of the Himalayas.

In the Middle Ages, the Arabs, quicker than the Europeans to set out and explore distant lands, developed a similar theory. They supposed that the world's weather was divided into seven latitudinal zones, with the Middle East in the fourth, central zone and Baghdad at the very heart of everything. The further one travelled from Baghdad, the more extreme the climate and the character of the people. Europeans, for example, were regarded by writers such as Muhammad al-Idrisi as of low intelligence and lacking religion because of the prevailing cold in which they lived. In the fourteenth century the historian and historiographer Ibn-Khaldun took this theory further, stating that civilised society was impossible outside the central three climate bands, beyond which the inhabitants were more beast than human: 'Their qualities of character ... are close to those of dumb animals ... The reason for this is that their remoteness from being temperate produces in them a disposition and character similar to those of the dumb animals, and they become correspondingly remote from humanity.' That includes us, up here in the second band from the top.

The theories were advanced still further by the sixteenth-century French philosopher Jean Bodin, who, armed with greater geographical knowledge than his Greek and Arabic predecessors, split the earth into six climatic and cultural bands, three in each hemisphere. Between the equator and the tropics, people were physically weak because they couldn't contend with the heat, but mentally highly developed. Between the pole and sixty degrees latitude the northern people were physically impressive but mentally negligible, the cold preventing their mental advancement. Bodin cites the example of the war between England and France: England won the fighting because, being further north, they had the physical advantage, but the French used their superior mental agility to negotiate more favourable peace terms than they might have earned from a more cerebral victor.

Montesquieu followed in the eighteenth century with some

frighteningly bananas theories about the influence of the weather on human character. Among his suppositions was that people in hot countries were almost compelled into slavery by being naturally idle and hence required force to make them work. The heat also made these southern peoples more randy, meaning that they had little inclination to do much else than laze around in the sunshine after a bit of how's-your-father. This would never happen in Europe because the less fearsome temperatures made one naturally more inclined to undertake hard work as well as keeping a lid on excess friskiness. In addition, the sexes were most unequal in standing in hot countries because – get this – the heat made women's looks decline more quickly than those of men.

In the same century, David Hume pointed out the folly of this kind of thinking in his essay 'Of National Characters', but even he still regularly couched his theories in the kind of language that would today earn him an opinion column in a national tabloid with an angry, finger-jabbing byline photograph. Athens and Thebes, he reasoned, were practically neighbours existing in the same climate, yet the Athenians were blessed with 'ingenuity, politeness and gaiety' while the Thebans were known for their 'dullness, rusticity and phlegmatic temper'. He didn't have much time for the modern Greeks, using them as an example of why climatic determinism was largely bunkum. 'The ingenuity, gravity and bravery of the Turks,' he noted, were in stark contrast to 'the deceit, levity and cowardice of the modern Greeks'. He also noted how 'the ingenuity, industry and activity of the ancient Greeks have nothing in common with the stupidity and indolence of the present inhabitants of those regions'.

National diasporas like that of the Jews, Hume said, preserved a people's characteristics despite centuries of wandering through all sorts of different climatic zones. When the European powers went off conquering and settling distant lands they preserved the national characteristics of their own area without being reduced to the feckless priapic indolence of the indigenous people.

He also believed that north Europeans fortified themselves against the weather with 'wine and distilled spirits' while the heat of the sun in the south heated the blood and 'exalts the passion of the sexes'. In Hume's world, north Europeans were a lot of drunks while their southern counterparts were a lazy bunch of shaggers.

Thankfully we moved away from such crackpot theories and the story of the weather and us shifted to more technical matters. For me, the way into them involved chocolate biscuits, a nice old lady and a beautiful piece of craftsmanship.

8

'WE LIVE SUBMERGED AT THE BOTTOM OF AN OCEAN OF AIR': THE BAROMETER AND THE CLOUDS

When I was little we lived next door to an old lady called Dolly. A widow, she lived alone in possibly the most typical old lady's house in the world. There were lace antimacassars on the armchairs and china dolls on the mantelpiece. The house smelled of lavender and bleach. She was tiny and frail, wore a shawl, had a blue rinse and favoured horn-rimmed old lady's glasses. You'd always get a birthday card from Dolly, usually a floral affair with an icky quatrain and 'love, Auntie Dolly' in slightly shaky handwriting inside. There would always be a pound note too. Even though we lived on a busy road Dolly's house was quiet, the ticking of the clock on her mantelpiece the only clue that time hadn't actually ground to a halt. Dolly would shuffle around in her carpet slippers and had a little black poodle called Penny whose yapping I can still hear from over the fence as I tore around our little garden wrecking plants with a plastic football.

Being a small boy I'd presumed that Dolly had always been old, so it was only years later that I realised that the beautiful young woman in the 1930s ball gown smiling next to the handsome young man in the tuxedo in the picture frame on her hall table was actually her with her late husband. What used to fascinate me when I went next door to Dolly's, usually to say thank you for a birthday card or to see if she wanted any errands from the shop, was the curious device hanging on her wall above and to the left of the photograph. It always baffled me. It looked like a clock encased in dark, carved, polished wood, only I couldn't tell the time by it. It wasn't ticking like the clock on the mantelpiece. It also looked a bit like the thermometer that would be stuck under my tongue whenever I was poorly.

One day, while Dolly was in the kitchen, I wheeled a pouffe out from the living room, placed it under the strange wall decoration, climbed on to it and had a closer look. It had hands like a clock, but there were words where the numbers should have been, words that said stormy, much rain, rain, change, fair, set fair and very dry. It was like a strange weather clock.

I heard the muffled shooshing of Dolly's slippers on the carpet as she shuffled from the kitchen carrying a tray with a doily, a glass of weak orange squash and a small floral plate with three chocolate biscuits on it.

'What's this, Auntie Dolly?' I asked, spludging a smeary fingerprint on the spotless glass. She carefully put the tray down on the hall table and stood next to me.

'That's my barometer, dear,' she said. 'It tells me what the weather's like.'

Dolly's voice was always clear and precise. Thinking about it now, she must have been quite well-to-do in her day.

'Bar-o-me-ter,' I repeated slowly.

'What does it say?' she asked. 'Can you read that word at the end of the pointer there?'

'Fair,' I said. Easy.

'That's right. That means that the weather is fair today.'

'How does it know?'

'It measures the pressure of the air around us, how heavy it is. It can tell the weather from that,' she explained. 'It's very old and very precious.'

I climbed down. That orange squash wasn't going to drink itself, after all.

'Do all old ladies have a bar-o-me-ter?' I asked through a mouthful of biscuit when I'd sat down in one of her armchairs.

She smiled. It took a while for a smile to work its way across Dolly's face but it always got there in the end.

'A lot of us do, dear, yes,' she said.

So that was that. As far as I was concerned a barometer was a thing old ladies had on their walls to tell them what the weather was like even though they could have just looked out the window. Seemed a bit pointless, I thought as I munched my chocolate biscuit, dismissing at a stroke one of the most important scientific inventions of them all and the greatest step forward in meteorology since Aristotle felt the wind under his tunic and thought, Hmm . . .

The remarkable works of a first-century Greek philosopher known as Hero of Alexandria were first translated in the late sixteenth century and they proved to be dynamite. Nobody knows anything about him and if you didn't know any better you'd be forgiven for wondering whether his works were an elaborate hoax. He is credited with the invention of a vending machine that dispensed sacred water: when a coin or token was dropped through a hole it would tilt a pan that delivered a measure of water before the pan snapped up again, causing the coin to fall off. I also like the idea of a wind-powered pipe organ, but arguably the most significant item in his writings was what appeared to be a primitive steam engine. At the heart of this machine – which from the plans and diagrams seems almost workable – was the principle that when air is heated it expands, something that got sixteenth-century academia hot under the collar and hence expanded the air immediately above their heads.

Out of this scientific flurry emerged Galileo Galilei with his thermometer, or thermoscope. He discovered that if you placed the open end of a glass tube in water its level was drawn up depending on the warmth of the air inside compared to that outside the tube, something he changed by cupping his hands around its closed end. Galileo's work in this field stopped there, short of adding a grading system for the level of liquid (he was, after all, a busy man). It was to be Evangelista Torricelli who in 1643 took the discoveries of Galileo and others further and created the barometer.

Hailing from the town of Faenza, south-east of Bologna, at nineteen Torricelli went to Rome to study science under Benedetto Castelli, a Benedictine monk, leading mathematician and friend and pupil of Galileo himself. Later Torricelli would study with Galileo in Florence and aligned himself with the Copernican school that questioned whether the planets really revolved around the earth: a risky commitment, certainly publicly, as the Church considered this heresy. They didn't think much of Galileo's talk of vacuums either, declaring firmly that God would never allow such a nothingness to exist.

Heresy or not, Torricelli set to work on the most baffling aspect of the vacuum: how the level of water in the tube would never rise above thirty-three feet. Galileo, ailing at the end of his life, couldn't come up with an explanation but Torricelli deduced that it must have been the weight of the air on the water in the dish outside the tube that pushed the level inside upwards. Once the water in Galileo's pump reached the critical height it was no longer lighter than the weight of the air and could thus rise no further.

Further experimentation was required and, with glass tubes more than thirty-three feet high being hard to come by, Torricelli set about finding a denser liquid that would enable smaller apparatus to be used. He settled upon mercury as a workable alternative. He placed some of the liquid in a tube just over thirty inches long, heated the tube to force out the remaining air and create a vacuum, then had his assistant put his finger over the end, invert the tube into a dish of mercury, remove his finger and, hey presto, twenty-nine inches of mercury slid up the inside of the glass tube.

'I claim that the force which keeps the mercury from falling is external and that the force comes from outside the tube,' he wrote. 'On the surface of the mercury which is in the bowl rests the weight of a column of fifty miles of air. Is it a surprise that into the vessel, in which the mercury has no inclination and no repugnance, not even the slightest, to being there, it should enter and should rise in a column high enough to make equilibrium with the weight of the external air which forces it up?'

He wrote to a friend that he envisaged a device 'designed not just to make a vacuum but to make an instrument which will exhibit changes in the atmosphere, which is sometimes heavier and denser and at other times lighter and thinner'.

After further experiments he was able to conclude that 'we are living submerged at the bottom of an ocean of air'. It was 1643 and Evangelista Torricelli had just invented the barometer. His weather credentials don't end there either: further musing on air pressure led him to finally subvert Aristotle's theories of exhalations by deducing that 'winds are produced by differences of air temperature, and hence density, between two regions of the earth'.

Torricelli lived for another four years before succumbing to typhoid at the age of just thirty-nine. His remarkable achievements – as well as his work with vacuums he also made massive contributions to the world of mathematics and was a craftsman lens-maker – make it baffling that he's not as well known as he perhaps should be. There is an engraving of Torricelli that was published just after his death in which he looks like he's not long got up after a heavy night: his hair's all over the place, there are bags under his eyes and beneath the elaborately sculpted beard there's an Eeyore-ish demeanour. Pin-up he may not have been, but Torricelli is up there with A-listers of weather like Aristotle and Descartes. It's too trite just to say he invented the barometer: what he did was push open the gateway to those who would come after him, ending centuries of Aristotelian stagnation. He wasn't the only one, there were other scientists working in the same area, but Torricelli saw his mercury experiments as more than a method of creating a vacuum in order just to say, 'Oh look, a vacuum.' The conclusions he came to went much further than his contemporaries and paved the way for the future of both meteorology and weather forecasting.

Seven years after Torricelli's experiment, Descartes's friend Blaise Pascal decided to tinker with the Italian's findings by trying the same experiment with red wine. Being far less dense than mercury, this necessitated a tube forty-six feet high. You can't help thinking that all the future author of the *Pensées* was doing was constructing an elaborate drinking game. At the same time, in Magdeburg the scientist and statesman Otto von Guericke decided to construct an enormous water barometer. It was attached to his house and, at fifty feet high, the top of the tube extended above the height of his roof. He dropped a wooden doll into the tube to float on the surface of the water, and locals started gathering to see her pop up the tube from behind the house during periods of fine weather.

Robert Boyle, of eponymous law fame, brought Torricelli's barometer to London in the 1660s, where his colleague Robert Hooke added a wheel connected to a pointer on a dial to the device, enabling him to deduce much smaller variations in air pressure. He 'found it most certainly to predict cloudy and rainy weather and dry and clear weather when it riseth very high'. The barometer stayed largely in the science laboratory for the next century until wealthier Georgians began to position them in their houses as symbols of status that also implied they were up to speed with the latest developments in science. Craftsmen from Venice and Milan arrived in London and began to produce some beautiful examples: technical improvements combined with a steady supply to bring down prices to a level that the upper middle classes could afford. This meant that as the nine-teenth century approached there were more people around the country with the equipment to not just keep detailed weather records, but to keep consistent records using the same methods.

The pace of progress towards the weather and shipping forecasts we know today was quickened when one of these amateur
meteorologists came up with one of those true genius ideas, the
kind of idea that asks, 'Why on earth has no one thought of this
before?'

Luke Howard was born in London in 1772, the son of a
lamp-maker. When he was ten years old, and a pupil at the
Quaker school in Burford, Oxfordshire, studying classics and
developing a strong interest in science, there was an extraordinary weather event that would change his life for ever to the
benefit of the history of meteorology: 'Around mid-morning
on Pentecost, June 8th of 1783, in clear and calm weather, a
black haze of sand appeared to the north of the mountains. The
cloud was so extensive that in a short time it had spread over the
entire area and so thick that it caused darkness indoors. That
night, strong earthquakes and tremors occurred.'

This was an entry in the diary of a Lutheran priest in Iceland,
shortly after an enormous volcanic eruption to the north-west
of Reykjavik. A sixteen-mile fissure had opened and for the
next eight months lava, steam and ash would pour into the
atmosphere. Most years the ash cloud would have been taken
north to the Arctic, but in 1783 the winds were from the north-
west and so the gases and dust wafted towards Europe. Two days
later ash was falling in Norway and within a week a brown-
tinged fog had settled over Berlin, where one diarist recorded
that 'the sun was dull in shine and coloured as if it had been
soaked in blood'. The wind then turned northerly and westerly,
with the first recorded reference to unusual conditions in Britain
occurring on 22 June, two weeks after the first eruption,
with reports of a relentless gloom in Norfolk and a Hampshire

clergyman noting that 'blades of wheat in several fields are turned yellow and look as if scorched with frost'. The ash cloud would eventually reach as far as Syria and cause unseasonable frosts in central Asia. In Britain and Ireland, what became known as the 'dry fog' settled above the ground like a thick blanket and stayed for months. The death rate increased by an estimated twenty-three thousand in England alone, many of the fatalities agricultural workers breathing in the dust while working in the fields. The days were stiflingly hot and dry; the nights freezing-cold and frosty. Everything was still.

In the Cotswolds Luke Howard became transfixed by the spectacular sunrises and sunsets induced by the ash. Every evening he would watch the sun sink towards the hills, its great disc turning from dark orange to blood-red as it descended. Then, as it slipped from view, a light show began: bright green hues dancing around the heavens, turning purple and pink before dusk became night. The extraordinary conditions meant that the aurora borealis was regularly visible, and in mid-August a spectacular meteor burned its way across the evening sky. This spellbinding, choking summer provoked in the young Luke Howard a lifelong love for and devotion to meteorology.

On leaving school he was apprenticed to a pharmacist in Stockport before opening a pharmacy in London's Fleet Street in 1793. He would later set up a successful pharmaceutical business, but the heavens remained his true passion and as the nineteenth century dawned he began to keep meticulous weather observations, recording readings from his barometer and thermometer and writing descriptions of the conditions of the day. This he would continue to do for forty years.

Howard realised that while he could note down accurate

readings and figures from his instruments the written descriptions proved difficult, not least when it came to clouds. What frustrated him was the lack of a uniform language applied to their descriptions. Most people had always considered clouds to be unclassifiable: no two were the same, they were random, with different textures, thicknesses, consistencies, even colours. But Howard began to notice some distinct types. He wasn't the only one: in France in 1802 a former soldier turned biologist named Jean-Baptiste Lamarck suggested five classifications of clouds in his paper 'On Cloud Forms', stating that 'besides the individual and accidental forms of each cloud, it is clear that clouds have certain general forms which are not all dependent on chance but on a state of affairs which it would be useful to recognize and determine'.

Later that same year, Luke Howard presented a paper to the Askesian Society, a scientific debating club of which he was a founder member. It met in the basement of a building in London's Lombard Street and would often be a riotous affair: at several meetings members would inhale nitrous oxide and stumble around the place giggling. It was cold in the unheated room when Howard stood up to speak, but what the members huddled in their cloaks heard that night would change things for ever. A keen botanist, Howard was familiar with the Latin classifications that had been introduced to plants by Carl Linnaeus and proposed a similar system for clouds. In his paper, Howard suggested four basic types. There was *cirrus*, from the Latin for 'curl' or 'hair', which he classified as 'parallel, flexuous or diverging fibres extensible by increase in any or in all directions'. *Cumulus*, the Latin for 'pile', were 'convex or conical heaps, increasing upward from a horizontal base', while *stratus* was 'a widely extended, continuous, horizontal sheet, increasing from below upward', from the Latin meaning 'layer'. Finally, there was *nimbus* (from the Latin for rain), 'a cloud, or system of clouds from which rain is falling'. The four terms could be joined for greater accuracy, hence for example we get *cumulonimbus* for a large rain cloud.

That night in a room full of amateurs, half of whom were wondering if they'd ever get the nitrous oxide out, Luke Howard changed the way we talk about the weather. The genius is in its simplicity: the terms are so familiar to us now that it's hard to imagine the forehead-slapping wonderment that greeted their dissemination. What clinched it for Howard was couching the terms in Latin, the closest the academic and scientific worlds had to a universal language. As Howard wrote when he published his paper as *Essay on the Modification of Clouds*, Latin was 'an universal language, by means of which the intelligent of every country may convey to each other their ideas without the necessity of translation. And the more this facility of communication can be increased, by our adopting by consent uniform modes, terms, and measures for our observations, the sooner we shall arrive at a knowledge of the phenomena of the atmosphere in all parts of the globe, and carry the science to some degree of perfection.' Once his proposals had been published, Howard's fame grew. Goethe was a fan and the two men corresponded regularly. The German writer was himself a keen weather observer and used Howard's classifications in his journals. He wrote that Howard 'was the first to hold fast conceptually the airy and always changing form of clouds, to limit and fasten down the indefinite, the intangible and unattainable and give them appropriate names', and even wrote four poems dedicated to the amateur meteorologist who named the clouds. If it hadn't been for Howard addressing a chilly basement Shelley might not have written 'The Cloud'. Turner was an admirer too, producing a series of cloud studies directly inspired by Howard's work. Constable also acknowledged him and, through Goethe, the painter Caspar David Friedrich came to be inspired by Howard's naming of the clouds. Not a bad scientific and cultural legacy for a Fleet Street pharmacist.

Ironically, for a man so adept at finding words for the apparently unnameable, Luke Howard proves difficult to describe. For one thing, unlike just about all the other main players in the weather story he wasn't a professional scientist, philosopher or meteorologist. His was a strictly amateur career; he was a pharmacist and a businessman first, a meteorologist next. Indeed, his paper was in part only read that night because members of the Askesian Society had to deliver a paper at meetings or be fined. His *Essay*'s publication came about only because the editor of the *Philosophical Review* happened to be in the audience and asked if he could print it. That's not to suggest he would otherwise have kept it to himself, but this was one of those happy accidents where the timing and the material were exactly right and led to a course of events that changed the way we look at weather for ever.

I like to think that, like George James Symons and the rain, Luke Howard always retained the sense of wonder that he'd felt watching those ash-cloud sunsets as a boy, sitting in a Cotswold field picking at the grass and watching a celestial lightshow like no other before or since. One thing I particularly like about Luke Howard is that he wasn't motivated purely by the pursuit of knowledge: he loved the weather and loved the clouds and was awed by them. What he wanted was to enable new powers of description regarding the silent, majestic piles of white in the sky. Most of all, he saw the beauty in the weather. 'The sky too belongs to the Landscape,' he wrote. 'The ocean of air in which we live and move, in which the bolt of heaven is forged, and the fructifying rain condensed, can never be to the zealous Naturalist a subject of tame and unfeeling contemplation.'

Nonetheless, Howard and his fellow amateur weather recorders were quietly and busily tapping their barometers, checking their thermometers, squinting at the sky, classifying the clouds, retiring to their studies and scratching away in their weather journals. This dedicated army of enthusiasts, from members of the aristocracy to country parsons to army officers to a diligent Fleet Street pharmacist, didn't know it at the time but they were all vital nodes in a giant historical weather diagram. As the nineteenth century opened, the data and the instruments for reading the skies and the air were now in place, we had nearly all the required language for classification and the momentum was gathering towards a nationwide system of recording and indeed forecasting the weather.

So I was wrong. The barometer wasn't just a thing old ladies had on their walls that told them what they could see for themselves if they looked out of the window. As I stood on that pouffe in Dolly's hallway I was looking at one of the greatest inventions in history, a beautiful piece of both craftsmanship and workmanship kept carefully polished and dust-free by a kind old lady in south-east London. There was a direct line from Evangelista Torricelli to Dolly next door, but as a little boy with biscuit crumbs down his front I had no idea of its existence.

One day Dolly's house caught fire. Fortunately my mum happened to be at the sink, overlooking Dolly's house, and saw the flames leaping up the curtains in Dolly's back room. She threw down the washing-up brush and flew out of the front door with her hands dripping, calling out to my dad to phone the fire brigade. I ran out after her and caught up as she rat-a-tat-tatted on Dolly's front door and kept her finger pressed against the doorbell. We heard Penny barking. A shuffling noise approached and the door opened.

'Come on Dolly, come with me,' said Mum with a combination of sing-song kindliness and urgency. Dolly immediately did what she was told. I scooped up Penny and we walked Dolly slowly back to our house. When Mum had sat her down on the settee and explained what was going on, Dolly cried. I'd never seen an old lady cry before and I didn't like it, so I went back outside where the fire engines were just arriving. The fire had spread to the upper floor and smoke was pouring from between the roof tiles into the sky. I watched the firemen going through the front door wearing breathing apparatus and after a while the fire was out. When I went back into our house Dolly was still crying.

Dolly's house was too badly damaged for her to move back in so the council came and took her away to live in an old people's home. She died two weeks after moving there. After that some men came and cleared Dolly's things and I went in with my dad while they were there. The smell of lavender and bleach had gone; everything smelt smoky and damp. It was cold. Dolly's house was never cold. The wallpaper in the hall was peeling off and was streaked with brown water stains from the firemen's hoses. The picture of the smiling young man and woman had gone from the hall table but the barometer was still hanging there. The glass over the dial had smashed and there were water droplets inside it. The two pointers hung straight down. I wanted to take it home and see if Dad and I could fix it, but it wasn't mine to take so I didn't ask.

The next day a lorry came to take away a skip full of ruined furniture and bits of burnt wood. I'd peeped over the top of it earlier that morning and saw the smashed barometer sitting on the top, its fluted sides now scarred with chips and digs. I'd stood on my toes and strained really hard but I couldn't reach it. The skip was lifted on to the back of the lorry and it drove off up the road. I went back inside, where the radio in the kitchen was broadcasting the shipping forecast.

9

FRANCIS BEAUFORT HARNESSES THE WINDS

I am often asked why the shipping forecast has become such an integral part of our culture. Why has this litany of names, numbers and curious phrases broadcast at odd and frequently antisocial times of day become a much-loved tradition? After all, we're assailed by weather forecasts on national and regional television and radio, in the newspapers and on the internet, nearly all of which are easier to understand and more locally applicable than 'the ships'. The shipping forecast doesn't even cover those of us who aren't domiciled on remote, windswept islands or who aren't chugging around the choppy waters that surround us, a corncob pipe clenched between our teeth and the whiff of oil and brine in our nostrils.

One reason I think the shipping forecast is held in a different and higher esteem than the more conventional forecasts is that it speaks wholly of the sea. Other nations have their own shipping forecasts – France and Spain, for example – but none hold the same fascination among the populace as that broadcast by the BBC four times a day. Songs, poems and books have been written about it, and I think the reason for our unique cultural response is that unlike, say, France and Germany, we are entirely surrounded by the sea. It has helped to define our history and culture, and the small geographical size of our country (in which we are never more than seventy-four miles from the sea, wherever we may be in Britain or Ireland) keeps the sea constantly at the back of our mind.

These days we're not as dependent on the sea for security and trade as we once were, but we are still pretty dependent on the weather. The sea plays a crucial role in that too, and the story of the advances in weather knowledge and forecasting is also bound to the sea. It was the loss of ships for want of weather information that led to the creation of the Meteorological Office, and the grading of the winds, a key aspect of the shipping forecast, is down to a great seaman, an Irishman whose most lasting achievement was something he scribbled down in the cabin of his ship while at anchor one day in January 1806, which defined once and for all that most elusive and invisible of all the weather elements. He was Francis Beaufort, the father of the wind scale that bears his name. What Luke Howard did for the clouds, Beaufort did for the wind.

Appropriately for a man whose life would revolve around water, Francis Beaufort was born at the confluence of two rivers, the Boyle and the Blackwater, in Navan, County Meath. His grandfather, Daniel Cornelis de Beaufort, had arrived in Ireland as chaplain to Lord Harrington, the Viceroy of Ireland, and was given the benefice of Navan and pastoral care of Dublin's Huguenot cemetery. Daniel's son, also Daniel – and a multi-talented man who, among other things, wrote the topographical *Memoir of a Map of Ireland* in 1792 – married an heiress named Mary Waller, from Allenstown, County Meath, in 1767 and Francis was born on 27 May 1774. Curiously for a child whose early years were spent some distance from the sea, Francis was obsessed with it. His sister Louisa wrote early in the nineteenth century that by the age of five he 'had manifested the most decided preference for the sea' and looked destined for a naval career. An associate of Francis's father, a Captain Mayne of the Royal Navy at whose wedding Daniel Beaufort had officiated, had already suggested entering the boy for the Navy and offered to 'give him standing', a common practice by which the children of friends could be entered in a ship's book to clock up time at sea without actually leaving shore.

Daniel Beaufort may have appeared to have been a modest country parson, but he was far more than that: intellectually curious, he had always been a major part of Irish cultural life. His finances were always precarious, however, and the evasion of debts was the main reason for the family moving away from Ireland when Francis was six, first to Carmarthenshire and then Chepstow. The peripatetic nature of these early years meant that Francis and his siblings had no formal schooling beyond what their father and grandfather had been able to provide at home. In 1782 the family moved to Cheltenham and Francis and his brother William were enrolled in the local Hospital Free School. However, the experience lasted just two days before the headmaster sent both boys home, saying that he refused to teach them on account of their 'Hibernian accents'.

Within two years the Beauforts' financial worries had eased enough for them to return to Ireland. They settled in Dublin at a time when the city was at its cultural and economic peak, taking a house in Mecklenburg Street. Finally, at the age of ten,

Francis could embark on a settled, formal education, spending
three years at the city's Military and Marine Academy – run by
the magnificently named Master Bates – where Francis showed
great promise as a draughtsman. So much so that his father
trusted him at the age of fourteen to take latitude and longitude
measurements for his great Irish topographical opus.

A visit by his father to Captain Mayne brought the bad news
that Francis's name had not been entered in any ships' books
after all, which made a career in the Navy less appealing. Instead
it was decided that merchant shipping was the way forward (its
financial prospects were much better, for one thing) and Francis
was enrolled on the *Vansittart*, an East India Company ship out
of Gravesend. Francis made the long journey from Dublin to
Kent via Holyhead and on 17 March 1789 the *Vansittart* set sail.

Francis's first seven days on board ship were lashed by fearsome storms and he could barely leave his bunk for seasickness.
As he lay there sweating and shaking, his insides churning as the
ship juddered, swayed and rolled in the gales, the teenage midshipman could never have foreseen the contribution he would
make to the maritime and meteorological worlds. His first significant maritime memories would be of storms, of the howl of
the wind and the crashing of the sea, the creaking of timber and
the wrenching of his guts. It was, at least, an appropriately meteorological start to his career.

Once the storm had abated and he could emerge pasty-faced
from his bunk, Beaufort proved to be a popular member of the
crew. He took his initiation in good spirits, or as good as your
spirits can be when you're lashed high in the rigging until you
promise two gallons of beer to your new mess mates, and within
three weeks was a regular fixture on the officers' quarterdeck,

taking the noon altitude of the sun to fix the ship's position and taking moon and star sights at night.

His talents were further confirmed when the *Vansittart* was to put in at Batavia (now Jakarta) after a hair-raising journey around the Cape of Good Hope. It wasn't a scheduled stop but rations were low and members of the crew were showing signs of scurvy, so the Company's offices in the Dutch East Indies were a necessary diversion. Beaufort − on his first voyage, remember − by taking soundings and bearings was able to survey a quicker route in to Batavia through the Gaspar Strait than anyone before, and when the *Vansittart* was damaged in port, necessitating a longer stay than planned, Beaufort borrowed a sextant and recorded a more accurate bearing for Batavia than the Dutch had ever made themselves. The *Vansittart* was lost soon afterwards to a combination of running aground and subsequent looting and burning by pirates, once the crew had sailed for help in a flotilla of small boats.

The captain of the *Vansittart*, Lestock Wilson, had been impressed by Beaufort's skill and maturity on the voyage, and especially amid the trauma of losing the ship. When the *Vansittart's* crew eventually arrived back in London Wilson was given command of the *Exeter* and immediately invited Beaufort to join his crew. The only problem was the *Exeter* hadn't actually been built, and wouldn't be ready for another two years. He advised Beaufort to get a couple of years' experience in the Navy and rejoin his crew when the ship was ready. Daniel Beaufort, who seems to have been the most well-connected Irish country parson in history, tugged at the sleeve of an old friend from Trinity College, Dublin, whose father was now Treasurer of the Royal Household. Appropriate strings were manipulated and, a mere six weeks after his arriving back from the burnt-out hulk of the *Vansittart* on the other side of the world, Francis boarded the frigate HMS *Latona* as a midshipman. It was June 1790, the British and Spanish were giving each other the moody eye over sovereignty of the Nootka Sound off Vancouver Island and that busy sprite, the spectre of war, was lacing up its shoes and looking for its keys ready to preside over what looked likely to be the latest in a steady line of Ibero-British seaborne skirmishes.

The differences between the East India Company and the Royal Navy were immediate and obvious to Beaufort. The discipline was far stricter, for one thing: he witnessed his first flogging and when the *Latona* moored at Portsmouth as part of its patrolling of the western English Channel he was given the distasteful task of being part of a press gang sent to nab unsuspecting drunks and force them into a life as Jolly Jack Tar. There was nothing to do when out on patrol, other than board a couple of French merchant ships, and when the Spanish backed down over Nootka Sound the crew were paid off and sent home.

The next few years were largely uneventful, with Beaufort's commissions allowing him to explore ancient remains in the Mediterranean and take soundings where he could until he took part in the Battle of the Glorious First of June in 1794. He was promoted to Third Lieutenant and transferred to the *Phaeton*, one of the greatest frigates in the Navy. One of his first voyages was to escort the Earl and Countess of Elgin to Constantinople where the Earl was taking up a post as ambassador to the Ottoman Empire. On the way they put in at Palermo and Beaufort was introduced to Nelson, who was impressed by the young Irishman and noted Lord Elgin's glowing testimonials too. Beaufort was at last being noticed yet despite his obvious talent, successful career and admirers in high places he was racked with low self-esteem and bouts of depression that he called his 'blue devils'. On top of this he developed a skin complaint, most likely porphyria, which his diet, seawater and the constant Mediterranean sun did little to ease. Then, in the autumn of 1800, five years before he created his table of the winds, Beaufort was involved in an incident from which he was lucky to escape with his life.

The *Phaeton* had been cruising off the coast of southern Spain when the crew's suspicions were raised by a distant three-masted brig flying the Spanish colours. The ship put in to the port at Fuengirola and the *Phaeton* waited for her to re-emerge, thinking her ripe for plunder. When the ship failed to reappear Beaufort and a few men set out in a cutter under cover of darkness to investigate. A flare went up, illuminating the scene just as the boat was about to reach the Spanish ship, and the game was up: the Navy men came under fire. Despite this the detachment, led by Beaufort, still boarded the brig. As Beaufort climbed over the gunwale he was attacked by a man with a sabre and wounded in the head and arm. No sooner had he arrived on deck, already bleeding heavily, than he was shot at close range with a blunderbuss. Fortunately Beaufort had turned to one side as the gun was fired so the volley of lead somehow missed everything important. He was still badly wounded, losing two pints of blood and falling into a delirious fever that lasted for two days, days in which his survival was by no means certain, but thanks to the surgeons at the military hospital in Gibraltar every piece of lead shot bar one was removed:

Beaufort would carry this piece of shot in his lung for the rest of his life.

Despite his heroism during a successful mission – the captured ship *San Joseph* was refitted and put into service – Beaufort still felt snubbed by the naval hierarchy. He might have expected to be given command of the *San Joseph*, renamed HMS *Calpe*, in recognition of his role in capturing her, but the command was given instead to the close relative of someone high up at the Admiralty. Nepotism aside, Beaufort was still suffering the after effects of his injuries and was told to forget about returning to the service for a year at least, but he still took it as a sign that whatever he did he would always be an outsider. He was made nominal commander of a Portsmouth fire ship, HMS *Ferret*, but this was a post in name only and, given that he wasn't on active duty, he was placed on half pay for the duration of his recuperation, which made things financially interesting.

Such apparent nonchalance towards his career prospects on the part of the Admiralty allowed Beaufort's 'blue devils' into his mind again and, with little to keep him busy, he had too much time to think. He still had ambitions, and on the way back from his convalescence in Portugal he would look out at the flotilla in which his ship was returning and fantasise about capturing a fleet of the same size with a single ship. Yet at the same time he felt a curious emptiness, a sense of ennui that told him the glory of heroism was meaningless, to the extent that he seriously considered forsaking the Navy altogether for a simple life in the Irish countryside, 'bargaining for constant spring'. He arrived back in London in early September 1801 feeling apprehensive about the future. However, within hours of his return a sense of duty led him to present himself at the

Admiralty to ask for command of a ship. Despite his well-documented acts of bravery in Fuengirola, Beaufort was kept waiting for an incredible six hours and was then told by the First Lord of the Admiralty that, with others above him in the pecking order, he would be best advised to apply for a pension to compensate him for his wounds. His application for the pension was then rejected – by the First Lord of the Admiralty. The pension was eventually granted after Beaufort protested strongly at the farcical decision, but in early 1802 he went back to Ireland in low spirits.

Back in Collon, the Beaufort family home in Meath since 1790, the curious dichotomy of his ambition was perfectly represented. In contrast to the noise and smoke of battle, Beaufort revelled in the quiet of the Louth countryside. Curiously, the blood loss he'd suffered and the change of diet from that at sea had completely cured his porphyria, and he spent many contented months reading and discussing philosophy and science with the formidable Dublin intelligentsia of the day. Yet at the same time Beaufort, feeling rested and reinvigorated, still made regular trips to London to apply for the command of a ship. This was a long way from the Ryanair age: the round trip would have been a major undertaking of anything up to ten days. Beaufort certainly wasn't just dropping by to show his face; something inside was driving him repeatedly back to the Admiralty. He was repeatedly turned down too, but not necessarily through any bias against him or his abilities: there was no war with France at the time and there just wasn't the same call for commanders as there had been. Nonetheless, as he paced the corridors lined with portraits of naval men who had achieved just the kinds of things he hoped to achieve, and waited for a verdict that, deep down, he knew would be a refusal, Beaufort must have harboured a pain somewhere within him, the fear that he would never be truly accepted. There was a permanent sense of otherness defining Francis Beaufort, the feeling that he was always destined to be the outsider. Maybe he felt that to the Admiralty he was just the son of a parson from away across the Irish Sea, the one who'd been turned out of the Hospital School after two days for his 'Hibernian accent'. He didn't belong here, for all his bravery, injuries and clear brilliance at taking soundings and observations, but back in Collon he would always be the one home from the sea, the navy man with the grapeshot in his side, the one who'd been away for years and missed being a part of the rhythms of everyday life there. It seemed as if he was an outsider wherever he went.

His time in Collon wasn't spent moping, however. Daniel Beaufort, as an established member of Ireland's scientific circle, had become friendly with Richard Lovell Edgeworth, also a man of science and literature as well as the father of the novelist Maria Edgeworth. Francis came to look upon Edgeworth as a mentor; the older man fired the naval commander's love of learning and science and introduced him to a number of intellectual luminaries. It was around this time that Beaufort fell in love with Edgeworth's daughter Charlotte. She was fiercely intelligent, opinionated and beautiful, and Beaufort had fallen for her the first time he laid eyes on her. Although utterly smitten, his chronic lack of self-confidence meant that he kept his feelings entirely to himself, unaware that the popping hearts that appeared around his head whenever Charlotte was in the vicinity had given the game away long ago.

In October 1804, as he made ready for another journey to

the Admiralty, a letter arrived from Charlotte. She had clearly realised that the initiative would have to come from her and dispatched a letter that, although couched in the careful phrasing of the day, essentially said, 'Come and get me, Frankie.' While in England, Francis visited Charlotte's father, who happened to be in Oxford at the time. Buoyed by Charlotte's letter, he plucked up the courage to reveal his true feelings to Richard and moot the idea of a marriage proposal. Edgeworth's response was not what he'd hoped. The older man would have been unquestionably delighted to have Beaufort as a son-in-law but the girl was in poor health and, having lost his wife to tuberculosis, he didn't want the same heartbreaking destiny for someone he'd come to regard as a bit of a protégé. It was a low point for Beaufort. Seemingly thwarted in love, struggling financially and with no sign of a commission anywhere on the horizon, he remained in London with his blue devils for the next seven months.

In June 1805, however, his career fortunes improved unexpectedly, at least on paper, when a letter arrived from the Admiralty giving him command of HMS *Woolwich*. She was only a store ship so it wasn't by any means a glamorous commission, but Beaufort was in no position to turn her down.

At least she was a ship, if an unwieldy, battered one repaired with parts from other vessels. Even the guns were an assortment of leftovers, most of them entirely unsuitable for a ship like the *Woolwich*, and only enough to barely occupy half the gun ports anyway. Beaufort moved her from Deptford to Gravesend for further fitting and to take on stores ahead of a scheduled supply trip to India, but things didn't go well from the start. One of his crew fell overboard and was swept away, while the ship's boy fell

from the rigging and smashed his head on the deck. In Gravesend HMS *Porpoise* tied up alongside and elected to clear the
accumulated sludge from the hawsehole, the opening from
which the ship's anchor was raised and lowered. Unfortunately
the stinking, oozing gunk plopped straight on to the deck of
Beaufort's barge. A furious Francis complained to the Admiral,
who listened patiently before introducing the man who'd been
standing by the window throughout Beaufort's indignant
monologue about the failings of the other ship's crew, officers
and, in particular, captain. With the kind of fortune that only
seems to afflict lead characters in sitcoms, the man turned out
to be the captain of the *Porpoise*. Not only that, he was also the
man who would be Beaufort's commander on the forthcoming
voyage. Whoops.

For the next few months Beaufort watched great events pass
him by. The *Woolwich* sat waiting at Spithead as the autumn of
1805 turned into winter. News came through of the victory at
Trafalgar, tainted as it was by the death of Nelson. On 8
November he mustered his men, stood on the quarterdeck and
related the story of the victory and the death of England's greatest naval leader. It was a story of glory, regret and, for Beaufort,
what might have been. The combination of achievement and
deep sorrow reflected Beaufort's own psyche and he took the
death of Nelson badly. When he stood at Spithead on the deck
of his old jalopy of a stores ship, cobbled together from odds and
ends and equipped with leftovers and rejects, and watched the
Victory sail past with the other ships returning from Trafalgar,
Beaufort would have felt pride at what had been achieved
pricked with pangs of disappointment that he hadn't been there
himself.

It was a scene that summed up his naval career: he was close to greatness yet unable to be a part of or share it. While men with less experience came home with heroic tales and promotions guaranteed, Beaufort still waited for his orders to join the convoy heading for India. The wait grew so long that the *Woolwich*'s supplies had started to dwindle and she was sent across to the Isle of Wight to take on provisions under the supervision of Admiral Sir John Warren's fleet there. As 1805 became 1806 there was still no sign of departure, either with Warren or the India-bound fleet back at Spithead. Early on 13 January a signal went up for midshipmen to gather on Warren's flagship for orders, with a view to setting out for the West Indies. Beaufort's midshipman returned with news that while the rest of the fleet was to prepare to sail he had been given no direct orders for the *Woolwich*. Despite this, and the understanding that, although not directly expressed as an order, the *Woolwich* would eventually rejoin the Spithead fleet, Beaufort made ready to sail with all the vim and vigour of a man at last able to exercise the command granted to him months earlier. Warren's ships were to escort a convoy to the Caribbean and then lurk around the Atlantic for a while looking for French ships to irritate. While he would have preferred to be on a frigate, this was a much more appealing option to Beaufort than to sail east with the captain of the *Porpoise* giving him icy stares at every opportunity. Which is why, without receiving any orders to do so, Beaufort jauntily unfurled his sails and nosed out into the English Channel with the rising sun at his stern and, at last, a maritime horizon to head towards.

It is safe to say that Beaufort was in pretty good form that day. He'd been a little cheeky in sailing away with Warren

without orders, but then, no one had told him not to. He'd
spent his career so far being ignored and passed over by the
naval hierarchy and this time their offhand treatment prompted
him to snaffle an opportunity that had not been supposed to
come his way. That evening as the sun set over a meniscus horizon uninterrupted by wharves, buildings or moored ships,
Beaufort dined alone in his cabin. He was probably happier
that night than he'd been in years. He was on a ship of his own
as part of a fleet commanded by a man he admired and months
of adventure lay ahead. Add to that how his departure had got
one over on the Admiralty and you can almost picture Beaufort
sitting at his desk looking out at the sea as Portland Bill glided
past the porthole, slapping his thigh and letting out a giggle of
delight.

The legacy of that evening was to go far beyond the thought
of some loud exclamations echoing around an office at the
Admiralty. When he'd finished dinner Beaufort took his journal from his desk drawer, opened it at a fresh page and began to
write. He had no inkling at the time, but what Beaufort wrote
that night, in a journal that he'd accidentally started using upside
down, would be his legacy, the achievement for which he
would be remembered two centuries later and beyond, something that would help sailors and landlubbers alike, something
that would help save lives as part of the shipping forecast and
every weather forecast worth its salt.

At the top of the left-hand page he wrote: 'Hereafter I shall
estimate the force of the wind according to the following scale,
as nothing can convey a more uncertain idea of wind and
weather than the old expressions of moderate and cloudy etc
etc.'

Then, underneath, he wrote the following list:

0   Calm
1   Faint air, just not calm
2   Light airs
3   Light breeze
4   Gentle breeze
5   Moderate breeze
6   Fresh breeze
7   Gentle steady gale
8   Moderate gale
9   Brisk gale
10 Fresh gale
11 Hard gale
12 Hard gale with heavy gusts
13 Storm

After this, midway down the page, he wrote, 'and the weather as follows, etc':

B   Blue skies
f   Fair weather
d   Dry warm atmosphere
s   Sultry
p   Passing showers
c   Clear, that is clear hard horizon but not blue sky
cl   Cloudy
w   Watery sky
wd   Wild, forked, confused, threatening cloud
dk   Dark, heavy atmosphere

l	Lightning
t	Thunder
g	Gloomy dark weather
gr	Grey threatening appearance
h	Hazy
dp	Damp air
fg	Foggy
r	Rain
sr	Small rain
dr	Drizzling rain
hr	Hard rain
sh	Showers
hsh	Hard showers
sld	Settled weather
sy	Steady breeze
sq	Squally
hsq	Hard squally
bl	Bleak horizon and clouds
thr	Threatening appearance

The neat hand and careful graduations suggest that this hadn't been written on a whim; the scale hadn't just occurred to Beaufort as he sat chuckling to himself in a postprandial haze, feeling the ship swaying in the water and hearing the waves washing past. It was written in the journal that recorded the weather on the voyage and, given that 13 January was the first day at sea, it was logical for him to write it up on the day the *Woolwich* departed. The scale was clearly something that had been on his mind, something he'd intended to use when he next put to sea. In fact, strictly speaking he hadn't even devised the list himself. Beaufort would have seen the scale devised by Alexander Dalrymple that featured in his *Memoir* of 1779. Dalrymple, the Navy's first ever hydrographer, was an obsessive producer of charts designed to confirm the existence of the great undiscovered continent of Terra Australis in the southern Pacific. (Indeed, it was largely because of Dalrymple's work that Captain James Cook 'discovered' Australia.) Before his Admiralty posting, however, Dalrymple had been hydrographer to the East India Company, Beaufort's first employer. Dalrymple had set himself the task of defining the best times of the year for the East India ships to sail in order to be ahead of the competition, squirrelling himself away with piles of ships' journals in order to discover the patterns of wind and weather most favourable to the East Indiamen. His task was made difficult because the descriptions of the wind strengths in particular were so arbitrary, which led him to compile a list of his own terms, with each assigned a number on a scale from zero to twelve.

Dalrymple was appointed hydrographer to the Navy in 1795 and expected the same level of cooperation from the ships' captains there, but he was to be disappointed: his office received only one report in three years, and that was just someone messing about. Thankfully for Dalrymple, and indeed the shipping forecasts of the future, Beaufort was an extraordinary exception. The two men had met through Lestock Wilson, the captain of the *Vansittart* and by then a highly successful ship owner, a few weeks after Beaufort had been given command of the *Woolwich*. Dalrymple recognised a kindred spirit, especially when Beaufort told him that he'd kept weather journals from his earliest voyages. He provided Beaufort with copies of his charts and arranged for him to have a new and rare boxed chronometer on board the *Woolwich*. Also among the papers and charts that Dalrymple showered upon Beaufort was a copy of his *Memoir*, the small pamphlet he had prepared while at the East India Company and which, crucially, contained his twelve-point division of the wind.

But even then, the idea had not been Dalrymple's. In 1759 the pioneering engineer John Smeaton gave a paper to the Royal Society about the weather and its relation to windmills. If, Smeaton posited, millers could calibrate the force of the wind it would be easier for them to know which sails to use for maximum efficiency. He had devised a type of anemometer that measured the speed of the wind and allowed millers to apply it to a scale he'd drawn up. Smeaton's eight-point scale included instructions to millers such as '4, fresh, wind heard against solid objects and agitation to trees, mill from 13 to 18 turns a minute'. Dalrymple had met Smeaton at the Royal Society in 1771 and the two became close friends: Dalrymple developed Smeaton's scale for the sea and this is what led to Beaufort sitting in his cabin that evening, modifying the work of his predecessors into something much more like the scale we know today. In fact, there's every chance that Beaufort was writing out the scale when the light on the Eddystone Lighthouse was visible from his cabin: a tower designed by Smeaton and first lit in 1759.

The history of wind definition goes back even further, to the early eighteenth century and Daniel Defoe who, after the Great Storm of 1703, devised a scale that went from zero to eleven (the Nigel Tufnel scale, if you like). Thus while Beaufort may not have come up with the idea, he tinkered with and improved existing scales and would set about making them standard, a process that began that evening in January 1806.

There are many adjectives that effectively describe Beaufort's achievements and personality but 'lucky' isn't one of them. His contented reverie after sneaking on to the mission to the West Indies lasted only three days: a mighty storm blew up in the eastern Atlantic and the fleet was forced to return to Spithead where Beaufort had to face the music. It began a miserable period for him: an unhappy voyage to the East Indies with commanders he didn't like, then some tedious supply missions before returning to Britain in March 1807 to find that, once again, the promotion he had expected had not come about – those with better family and social connections were still being placed ahead in the queue when it came to the handing out of gongs. Then on 7 April 1807 Charlotte Edgeworth died. Beaufort rarely left his cabin after this, even as the *Woolwich* was converted around him from a stores ship to a frigate and back again to a stores ship when the Admiralty changed their minds. In the spring of 1808 he emerged from his shell thanks to his increasing friendship with Dalrymple, only for this kindred spirit to die a few weeks later. When his *Woolwich* commission ended in 1809 he was gloomy about the future and beset by blue devils once more. So, when he was finally given the kind of commission of which he'd always dreamed – the eighteen-gun HMS *Blossom* – he was less than ecstatic, having apparently lost all his ambition and his stomach for a fight. He wrote to his brother William, 'my soul sickens at seeing all the world cutting or endeavouring to cut each other's throats, for what, they know not'.

His career looked up with the appointment of Charles Philip York as First Lord of the Admiralty, a man who valued Beaufort more than his predecessor, and Beaufort was immediately promoted to post-captain and given command of a thirty-two-gun frigate, HMS *Frederiksteen*. At around the same time he proposed to his old friend Lestock Wilson's daughter Alicia, telling his father in a letter that 'she is no beauty but has a good though delicate figure' – the old romantic – and soon after taking command of the *Frederiksteen* he set about refining his wind scale to include advice on the best setting of the sails for each grading of the scale.

He was approaching the end of his career at sea and he spent most of his final travels on the *Frederiksteen* happily taking soundings and visiting ancient ruins around southern Turkey and Syria. A skirmish with local bandits saw a midshipman killed and Beaufort taking a bullet in his side, but otherwise his final voyages were uneventful and he returned the *Frederiksteen* to Deptford where she was broken up in 1812. He spent the next few years working on his charts and writing up his maritime travels as a memoir. Alicia gave birth to two sons, he was elected to the Royal Society and in 1829, and after many disappointments and some financially uncertain years, he was appointed hydrographer to the Navy. Finally Beaufort's brilliance had been recognised; finally he was in a post where his talents could be realised. And, most of all, he could at last set about making his wind scale the standard measurement used by all sea captains.

The first captain to do so would be the man who, more than any other, is the hero of this story.

10

ROBERT FITZROY AND THE ORIGINS
OF THE WEATHER FORECAST

It's hard to believe it these days, but Upper Norwood was once a rural refuge of peace and quiet well away from the noise and bustle of London. Now completely subsumed in the suburban sprawl south of the metropolis, the only clue to its previous salubrious past lies in some of the architecture: the main roads are lined with big, solid four-storey Victorian houses, nearly all of which have been converted into flats.

I am standing in a wedge-shaped churchyard where Beulah Hill and Church Road come together and trying in vain to picture how Upper Norwood would have looked a century and a half earlier, before the minicab offices, before the takeaways, before the multiple entryphones and before the huge green plastic bins with numbers sloshed on them in white paint.

Norwood takes its name from the great forest that used to stand here. This south London ridge, its views north and south the main attraction for the rich and gentrified who came to live here, was once covered in mighty oak trees that were the timber of choice of generations of British shipbuilders: by all accounts the *Golden Hind* was constructed from trees felled in the Great North Wood. Hermits and gypsies inhabited the woods until finally squeezed out by a combination of the Enclosure Acts and deforestation, but some still remained in the early nineteenth century when the toffs began to arrive in search of peace and nice views. Felix Mendelssohn visited twice as a guest of his friend Thomas Attwood, organist at St Paul's Cathedral and apparently once a pupil of Mozart, who lived on Beulah Hill. On the first occasion, in 1821, Mendelssohn was recuperating from an accident and seemed quite taken with the restorative powers of the Norwood breeze: 'This is Norwood, famous for its good air,' he wrote to a friend. 'I have had a walk of two miles today and the air has really had a very salutary effect on me; in the three days that I have been here I can feel how much stronger and healthier I have become.'

Nobody would take the air in Upper Norwood purely for the benefit of their strength and health these days. I certainly hadn't as I stood in the churchyard of All Saints Church one blustery afternoon: a plumber's van pulling away from the traffic lights left a cloud of black exhaust drifting towards me, and the skeletal sentinel of Croydon's television tower rose into the grey, scudding sky nearby. The pulchritude and restorative air of Upper Norwood have, it's safe to say, been much reduced since Mendelssohn's time.

The person I'd come to see had taken a second home here towards the end of his life. He was a busy, driven man of incredible talent and ingenuity fuelled by a selfless desire to benefit the public good, but also a man plagued, like Francis Beaufort, by self-doubt and depression. It was this combination of drive and demons that had led to his moving out here on doctor's orders, as it was hoped he'd find calm and peace, and be able to return to his duties reinvigorated, even allowing for his advancing years.

The person I'd come to see was, appropriately enough, introduced to me by the shipping forecast. One of the sea areas has borne his name since 2002 (it's the only sea area to be named after a person), but it was only when I went beyond the name and into the world behind it that I came to know and admire a truly remarkable man. This whole weather story revolves around him, in fact, yet for a number of reasons it's not his contribution to meteorology for which he's best known.

The person I'd come to see was Robert FitzRoy. He had died nearly 150 years earlier, in a house I'd passed on my way to the church from the station, the one with the sun-faded red front door and the little green commemorative plaque on the wall. He'd died searching for the peace that life had never granted him and now, as I stood at the foot of his grave on a blustery afternoon, it was clear that he had not found it here in death, thirty feet or so from a busy road with the roar of traffic, the screech of sirens and the thunder of aircraft beginning their final approach to London's airports.

I stood grasping the iron railing that surrounds the grave. In front of me was the small concrete tablet placed there by the Met Office when the grave was restored from an unforgivably shabby condition in the early eighties. It displays a well-chosen quote from Ecclesiastes that I knew almost by heart from previous visits: 'The wind goeth toward the south, and turneth about unto the north; It whirleth about continually, and the wind returneth again, according to his circuits.'

A large, austere headstone faced me from the head of the grave, topped by a relief of an anchor propped against a cross. The stone is dark with hints of mossy streaks and I tried to imagine what it must have been like on the day in May 1865 that he was laid to rest here. It wasn't, by all accounts, the kind of send-off to which he was entitled: the nature of his death had seen to that. It was a sparsely attended, grim event according to one attendee, who wrote afterwards to Charles Darwin, a man whom FitzRoy had known intimately.

'I came back Thursday night hoping to be in time for the funeral,' wrote Bartholomew James Sulivan who had served under FitzRoy on the *Beagle* voyages that made Darwin's name and had worked with him at the Board of Trade. 'I found no one knew anything about it at the Board of Admiralty and I should not have known had not Mrs FitzRoy's mother come to his office to lock up his papers and they told me it was early on Saturday at the Norwood Church close to his house, and they had told no one but relatives as they thought under the circumstances it ought to be very private.

'It was a very quiet and plain funeral, just what I think all funerals should be,' he continued. 'Poor Mrs. FitzRoy would go and the two daughters were with her. We all waited outside and walked after her carriage and the same back, the brothers only going into the house. It was a trying scene at the grave. Poor Mrs. F. and the girls looked dreadfully ill and Mrs. F. gave way very much.'

I gripped the railings and pursed my lips at the thought of such an ignominious final chapter of a remarkable life. Robert FitzRoy deserved a funeral of pomp and decorum; he deserved long and glowing obituaries, people lining the streets as he passed, before a service attended by the great and the good, not half a dozen ashen-faced relatives and close confidants in a remote churchyard, watching his plain black coffin being lowered into the ground and hoping to get it all over with as quickly as possible.

The wind mentioned on the Met Office stone had helped to take FitzRoy to the far ends of the earth to achieve great things and leave a genuine, world-changing scientific legacy behind him, albeit one that caused him tremendous anguish and doubt in his later years. Once his career at sea was over he had thrown himself into the documenting of the winds so that sailors might depart with foreknowledge of the conditions into which they were about to sail. FitzRoy was directly responsible for saving countless lives, yet he was never appreciated in his own lifetime and has, in my opinion, been unfairly neglected ever since.

Robert FitzRoy deserves to be far better celebrated than he is. What we owe to him is incalculable, not just because, without him, Darwin would never have written *On the Origin of Species*, but also because he is the father of the weather forecast. Through tireless, selfless work FitzRoy made the world a safer place simply by introducing all the elements of the weather forecasts we take for granted today.

I once sat on a Galician peninsula looking out at the sunset over the patch of sea that bears his name in the shipping forecast, while on the bicentenary of his birth in 2005 I'd helped to organise a small memorial service here at his graveside because I believe that this brilliant, intelligent, flawed man is one of the greatest we've ever had. His life story is remarkable and ultimately tragic. His death was greeted with shuffling, coughing and swift changes of subject because he'd committed the final sin of taking his own life, a decision he had come to as a result of severe depression, from which he suffered terribly and which made his achievements all the more remarkable.

There is almost too much to say about Robert FitzRoy's life and achievements so I'll limit myself mainly to his last years; the period when he was most concerned with the weather. Of aristocratic descent, FitzRoy had a distinguished naval career after entering the Royal Naval College in Portsmouth at the age of twelve. He was the first person ever to score 100 per cent in the lieutenant's examination and took command of the barque HMS *Beagle* at the remarkably young age of twenty-three. He was a popular captain who commanded tremendous respect among the men serving under him.

In 1831 he was all set to take the *Beagle* on another survey-ing mission to South America. Although still in his mid-twenties, FitzRoy was already aware of the bouts of depression that would regularly cast dark clouds over his life. In addition, his uncle Viscount Castlereagh had cut his own throat in 1822 after experiencing paranoid delusions, while FitzRoy had in 1829 taken over the captaincy of the *Beagle* in South America after the previous captain, Pringle Stokes, committed suicide by shooting himself in the cabin FitzRoy would then occupy. Conscious that the necessary aloofness of a captain from his crew might not be beneficial to his mental health on a voyage that would take years, FitzRoy sought a companion to share his facilities and keep him mentally exercised. He solicited the help of his friend and mentor Francis Beaufort, via whom Charles Darwin, a student naturalist, was invited and agreed to go. What turned out to be a five-year voyage provided Darwin with all the evidence he needed to posit his theory of evolution, something

that would later bring him into sharp conflict with his erstwhile
cabin mate and friend: FitzRoy was increasingly pious and a
fierce creationist.

On his return FitzRoy was commended for the brilliance of
the charts he'd compiled, documents that would save countless
lives by giving accurate warning of dangerous terrain, and
awarded the Royal Geographical Society's Gold Medal in 1837.
Next he was elected Tory MP for Durham, and at the end of
1843 was appointed Governor of New Zealand. He'd stopped
there on the *Beagle* voyage and found the place in utter disarray,
without effective colonial leadership and rife with disputes
between the administration and the populace: as colonial post-
ings went, New Zealand wouldn't have made most people's
wish lists. Despite making the best of it under extreme, almost
impossible pressure, his leniency towards the indigenous people
caused grumblings in the corridors of the Colonial Office and
he was recalled in 1845 amid unfairly critical reports of his
tenure.

He arrived back in London in January 1846, forty years old,
married with four children under ten, effectively unemployed
and facing financial difficulties, having paid for a large number
of public works out of his own pocket in the reasonable but
mistaken belief that he'd be reimbursed. FitzRoy was tirelessly
dedicated to public service and regarded it as his vocation;
unfortunately not everyone shared his selflessness in this regard.
He was a man of boundless enthusiasm – one account by an
admirer recalled, 'I never in my life met a man who would
endure nearly so great a share of fatigue. He worked incessantly
and when apparently not employed he was thinking' – blighted
by his intense periods of chronically low self-esteem. When

returning from New Zealand he felt, rightly, that he had been a pawn in a wider game of political chess.

FitzRoy knew that, despite the career setback, he still had much to offer and, again in an echo of Beaufort, he hoped most of all for the command of a ship. A keen extoller of the virtues of steam power in shipping, he was appointed to oversee the construction and fitting out of HMS *Arrogant*, a hybrid ship that combined sail with a screw propeller powered by a steam engine, and the Navy's first ship to be powered by anything other than sail. FitzRoy set about his task with his customary zeal in order that the newfangled *Arrogant* should be seen as much more than a curiosity: indeed, she should become the finest vessel in the Navy. Once she was seaworthy, FitzRoy was appointed captain and, after some patrolling in home waters, the *Arrogant* was sent to Lisbon late in 1849.

Despite this being a posting to which he was perfectly suited, something happened in Lisbon that forced FitzRoy to resign his command. No one knows for sure, but it's apparent that, finding himself as the senior naval officer in the port, FitzRoy was required to undertake the entertainment of dignitaries and officials that cost more than his captain's wage could cover. When at the same time he received news from home that his wife was experiencing financial hardship it seems FitzRoy had a breakdown. It was never referred to directly, and there is merely an oblique reference to the necessity for a long rest for the benefit of his health in a curriculum vitae he wrote in 1852, but given his history of battles with mental illness it seems that in Portugal FitzRoy did succumb to severe depression. He would never put to sea in charge of a ship again.

Taking into account his experience in New Zealand as well

as a depressive period aboard the *Beagle* that had him writing to Beaufort that he was 'in the dumps ... ill and very unhappy' and had Darwin noting a 'morbid depression of spirits and a loss of all decision and resolution', it seems clear that FitzRoy suffered clinical depression for much of his life. He seemed almost to fear inactivity, throwing himself into whatever task he was given with superhuman effort to avoid the long periods of introspection that he believed could lead to the kind of consequences which compelled his uncle to cut his own throat.

It would be the best part of five years before FitzRoy would find a post in which his ideas and enthusiasm for public works would be exercised. The sudden death of his wife in 1852 would not have improved his mental condition, particularly as he had no particular large project to occupy his agile, fragile mind. Friends including Darwin had seen to it that FitzRoy was made a Fellow of the Royal Society in 1851 in an attempt to keep up his spirits, but his wife's death hit him hard and led to even more worry about the financial welfare of his four young daughters. The writing of his manual *Sailing Directions for South America* provided a brief distraction but it would be events in the late summer of 1853 that would change the course of FitzRoy's life and cement what should be his most important legacy.

At the end of August 1853 there was a conference of European nations in Brussels which was designed to coordinate a system of transatlantic weather observation and reporting for the benefit of shipping. The idea had come from America, where Matthew Maury, head of the United States Naval Observatory, suggested it would be beneficial for everyone to have a uniform system of weather reporting and observation.

After much lobbying the conference was arranged to take place at the residence of the Belgian interior minister.

Nine nations readily signed up for the conference but the British authorities were notably unenthusiastic – remarkable, especially as the Royal Society itself was vocally keen. The familiar frustration of bureaucracy lay behind this reticence: Maury's invitation had sat unanswered in London for months, arriving first at the Foreign Office before eventually being passed on to the Board of Trade, which in turn tugged it out from the bottom of the in tray and passed it on to the Marine Department. What each office in turn feared was signing up to something that was going to make a dent in their budget, and it wasn't until the very last minute that two men were dispatched to the continent – so late that they missed the entire first day – with firm instructions not to sign anything that involved financial commitment.

The conference was hailed by all parties as a great success. A uniform system was agreed upon and even the British delegation found itself inadvertently agreeing to, or at least recommending, the setting up of a dedicated meteorological department of some kind to record and share weather observations. It took months for the tangled squabbles between different departments to unravel themselves enough for Parliament to finally pass the funding for the setting up of a meteorological department. When, incidentally, the member for Carlow ventured that the collation of meteorological data could possibly lead to an assessment of what the weather might have in store over the ensuing twenty-four hours, the House collapsed into laughter. As if the weather could be predicted!

Fitzroy had already been sounded out about the possibility of heading any new department by Edward Sabine, then secretary of the Royal Society. FitzRoy had replied that he'd recommend the appointment of 'some neutral man who is interested in such subjects; whose character will guarantee his proceedings and give full time and thought to their pursuing', presumably while looking away, whistling innocently and pointing at a piece of paper on which he'd written his own name. There was probably little need for such heavy hints as Sabine, a man of considerable influence, was already convinced that FitzRoy was the person for the job. Indeed, on 1 August 1854, on Sabine's recommendation and almost a year after the Belgian conference, FitzRoy was appointed Meteorological Statist of the Meteorological Office of the Marine Department of the Board of Trade. Grand as the title may have been, the salary was a little more modest: five hundred pounds a year, plus half of FitzRoy's naval wage. When you consider that FitzRoy wasn't even given an office in which to work at first, leading him to take a room at his gentlemen's club instead, it's clear that the priorities of the establishment lay elsewhere. When you consider also that this was the birth of the Meteorological Office we know today, it's a startlingly low-key nativity.

In some ways this was not surprising, as by now both Britain and France had been pitched into the Crimean War. However, rather than obstruct FitzRoy's progress, the war came to underline just how important his new department could be: he'd been in the job less than four months when a huge storm set about destroying much of the Anglo-French fleet as it waited in the port of Balaclava, sinking a number of ships – including the famous French flagship, the *Henri IV* – and rendering others useless. It was a massive, not to mention expensive, setback, and

one that, as FitzRoy and his French counterpart Urbain Le
Verrier noted, could have been averted: the storm had begun in
the west and moved eastward along the Mediterranean. Nearly
all the damage could have been prevented if a warning of the
storm could have been sent to the ships at Balaclava using the
new technology of the telegraph.

FitzRoy's experience at sea, and his enthusiasm for barometers and thermometers, had taught him that the right instruments and knowledge of how to use them could also have
prevented the carnage, had they been in use on board the
vessels in port. He'd had first-hand experience, after all, on his
way back from New Zealand. He was aboard the *David
Malcolm*, and the ship dropped anchor at the western end of
the Strait of Magellan one night in April 1846. The captain, a
proper old seadog, eschewed modern nonsense like barometers and chose instead to rely on his own weather instincts
honed through years at sea.

It was a quiet, still evening when the captain prepared to
retire for the night but FitzRoy had consulted the barometers
he carried everywhere and noticed that the air pressure was
falling quickly. He expressed his concern that there could be a
storm on the way but the captain just looked out of the porthole at the mirror-watered stillness of the evening, waved away
FitzRoy's concerns and went to bed. At midnight the pressure
had fallen further and FitzRoy was convinced that some seriously bad weather was about to hit. He managed to persuade
the first officer to drop a second anchor and furl the sails and,
sure enough, within a couple of hours a storm was raging across
the harbour. The *David Malcolm* dragged slightly on one anchor
but remained clear of the nearby rocks against which she would

undoubtedly have been dashed had FitzRoy not cajoled the officer into disobeying his captain's orders.

FitzRoy was convinced that these potential disasters could be avoided if ships carried the right weather instruments. His task was made slightly easier when, on Christmas Eve 1854, he received news that he was to be given an office and a staff of three on Parliament Street, not far from the new Palace of Westminster that was then under construction.

Immediately FitzRoy set about commissioning hard-wearing, simple barometers that ships could carry, along with a manual that he'd written himself. He ensured the instructions were straightforward enough for any captain to understand – some of them were even in rhyming couplets for easier recall. In addition, FitzRoy's team plunged into the Admiralty records to annotate and interpret weather data from recently arrived ships' logs, while FitzRoy also persuaded correspondents around the country to send him their weather reports. He threw himself into redesigning the format of the weather logbooks agreed by the attendees of the Brussels conference – much to their indignation – and, having read Matthew Maury's North Atlantic weather charts from the United States Naval Observatory, set about making his own.

Here FitzRoy made his first major contribution to the science of meteorology. While he knew that the key to immediate improvements in safeguarding mariners was to provide them with the right instruments, he was clever enough to realise that stepping back and seeing the big picture was just as important in the long term. Hence FitzRoy adapted and improved on Maury's work by devising what he called 'wind stars'. He divided the Atlantic Ocean into a grid and placed in each square a visual representation of the winds mentioned in ships' logs over recent months: directional lines corresponding to points of the compass spread out from a central point; the longer the line, the more winds having been reported as coming from that direction. Lines were then drawn joining the tips of these wind lines together to make wonky star shapes, which would give captains at least an idea of where the winds were likely to come from in any given area, with circles inside the stars to denote how strong they'd been. With the division of the map to give weather information for each specific area of sea, the wind stars are clearly the direct predecessors of the shipping forecast. The only difference was that FitzRoy's charts weren't predicting the weather, only laying out its recent history, but at least mariners would have a rough idea of what to expect.

He also plotted what he called 'synoptic' charts (another of his contributions to the language of meteorology), demonstrating his theory that the winds and weather behaved in a vaguely regular way. Given the limited resources at his disposal, this was amazing stuff: FitzRoy's department was already making a difference in saving lives and they hadn't even got around to forecasting the weather yet. His small team beavering away at the weather data arriving at their office looked about as far from weather forecasters and life savers as you could get. On the outside they were no different from any other room full of middle-ranking civil servants ploughing methodically through piles of paperwork. But, underfunded and unappreciated, FitzRoy's bureaucratic underdogs were performing some of the most important work in the history of meteorology.

FitzRoy wanted to introduce a smaller, cheaper version of his barometer that modest fishing vessels – the ships that were most in danger from storms and towering seas – could use. For all his idealistic zeal, however, he had to concede these were impractical: fishing captains couldn't afford them, and anyway, they were generally far too busy to keep checking and recording barometer readings. Instead, FitzRoy pressed ahead with a plan to distribute barometers to the ten fishing ports most susceptible to storms and casualties, nine of them in Scotland and the tenth St Ives in Cornwall. The fishing fleets could consult the barometer readings before setting off and hence potentially avoid sailing right into a major storm. Fired by his concern for the welfare of those at sea, FitzRoy rather liberally interpreted a letter from the head of the Board of Trade 'agreeing in principle' to the idea as the green light, and immediately fell foul of the Board's accountants – not for the first time.

FitzRoy's actions may have seemed impetuous and dismissive of authority – he was already far exceeding the original remit of his office as essentially a statistics processing plant – but in all his years in the Navy he'd seen sailors and even whole ships lost for want of decent, reliable weather information and every storm, every ship posted as missing, every drowning of a fisherman bothered him more and more. The real-world reminder that the bureaucrats and bean counters didn't share his altruistic at-all-costs idealism angered and frustrated FitzRoy, as did the technical limitations of his pioneering work and techniques.

Nothing brought this more sharply into focus than one of the worst and most tragic weather disasters ever to occur on these islands.

11

THE BALLAD OF ISAAC LEWIS: THE *ROYAL CHARTER* STORM

Many people would claim the *Royal Charter* was an unlucky ship from the start. The company that commissioned her had gone bankrupt before she was even completed and at her launch in 1855 she had refused to budge from the slipway at the Sandycroft Ironworks on the River Dee, traditionally a sign of ill fortune for seafarers. When she ran aground straight after her launch it was almost as if she was trying not to go to sea at all. Then an error in ballast calculation almost saw her come a cropper on her first commercial voyage, necessitating an embarrassing unscheduled stop at Plymouth to put the problem right. The omens, particularly in such a superstitious world as the maritime one, were not auspicious.

Despite these early setbacks much was expected of the *Royal Charter*. She was built as a clipper, always the fleet-footed sprinters of the open seas, but with an iron–clad hull and the addition of a steam engine powering two screw propellers for when the wind dropped and her sails hung limp. She was a ship for all conditions, designed to work the long and busy run between Liverpool and Melbourne that needed a fast and reliable vessel: she could make the journey halfway around the world in sixty days, shaving a month from the previous length of passage.

It's unclear when Isaac Lewis first joined the ship, but the young man from the tiny coastal village of Moelfre in Anglesey would have found it an exciting opportunity. He came from a family of mariners – his father John was a seaman, as was his older brother William, twelve years his senior – and like many young Moelfre men he had grown up sitting on the shore watching the busy traffic of ships on their way to and from the Mersey. As a boy, Isaac would have dreamed of travelling the world as a sailor. After all, Moelfre's very existence revolved around the sea, and if you didn't seek employment in the nearby quarry that's where you looked for your living. A job on the *Royal Charter* would have fulfilled every ambition for the young man as well as endorsing his abilities as a seaman: as one of the most glamorous and modern ships on the sea, the *Royal Charter* would have had the pick of the best sailors in Britain.

Isaac Lewis had just turned twenty-one in August 1859, when the *Royal Charter* left Melbourne carrying around 370 passengers and approximately 120 crew. She was well under capacity – the ship could accommodate up to six hundred passengers – and hence was expecting to make good time. Many of those aboard had made their fortune in gold prospecting and were returning home with their future security carried about their person. The cargo hold also contained a massive amount of gold. The passengers were very much of their time: adventurers and prospectors exploiting the opening up of distant colonies. One man had originally been sentenced to seven years' transportation: he'd stayed on in Australia, struck lucky in the goldfields and had sent word that he was coming back with a fortune that would ensure nobody in his family would have to work again. Others had departed for Australia because the wages for skilled labourers were much higher on the other side of the world, and were returning home with their earnings. There was a sea captain cadging a lift after his own ship was lost off Fiji and a farmer who'd emigrated to New Zealand travelling over to Britain to buy agricultural machinery. From clergymen to miners, the passenger list of the *Royal Charter* represented a cross-section of Victorian society.

Under Captain Thomas Taylor, an old hand whose strict disciplinary regime made sure the job was done with minimal fuss, the *Royal Charter* left Melbourne on 26 August 1859. She made great time – but for some headwinds in the north Atlantic, the captain had estimated he'd have made the journey in an astonishing forty-seven days – and by 24 October had weighed anchor off Queenstown in County Cork. Fourteen passengers disembarked on to the port tender; one man, a brother of the MP for Meath, deciding at the last minute he'd join his departing friends after all and leaping aboard with his suitcase and coat-tails flying just as she was about to pull away. Three hours later, the *Royal Charter* departed on the final short leg of the journey to Liverpool.

The next morning dawned grey with a calm sea and a light south-easterly breeze. During the morning the wind veered to the north and increased, meaning the *Royal Charter*'s sails were furled and the engines employed, but she was still on course to complete the voyage in a very impressive fifty-nine days. In the middle of the afternoon there were notable wind increases in the English Channel: coastal villages in Devon and Cornwall suffered structural damage but the barometers on the *Royal Charter* registered nothing untoward, and as she approached Anglesey Isaac Lewis and the other seamen began to stow the sails that wouldn't be needed for the remainder of the voyage. As she passed Holyhead the passengers could see Isambard Kingdom Brunel's famous new *Great Eastern* there at anchor, where she was moored while undergoing sea trials. Meanwhile, barometer levels across Wales and the west were sinking dramatically as the wind chased and whirled around the rapidly diminishing pressure heading up the Irish Sea, but there was still no hint that the winds from south-western England were on their way: the barometers on board were still giving no indication of what was chasing them along the Welsh coast.

By eight o'clock that night, however, the storm had reached Anglesey and caught up with the *Royal Charter*, with winds reaching force 10 on the Beaufort scale. When she rounded the north-western tip of Wales and was off Point Lynas on the north coast of Anglesey Captain Taylor sent up flares to summon the Liverpool pilot to guide her into the Mersey, but the conditions had become so bad the pilot couldn't even leave port. The sea was getting up a tremendous swell and the wind screamed through the *Royal Charter*'s rigging. She pitched and tossed, waves breaking over her bow, the wind flinging spray across her decks, and an uneasy feeling ran through the passengers and crew. The wind shifted from northerly to north-north-east and raced up to hurricane force 12, something that began to push the clipper steadily from open sea towards the coast. Rowland Hughes, the captain of the lifeboat station just a few miles from where the *Royal Charter* was being battered by the storm, couldn't even put out to sea: 'The sea upon that beach,' he'd say later, 'was like nothing I'd ever seen before in my life.'

At eleven o'clock Captain Taylor dropped anchor and ordered the masts to be cut down. Two hours later the port anchor chain snapped, followed an hour later by the starboard one. The *Royal Charter*, her passengers and her crew were now at the mercy of the storm, the ship's steam engine and twin propellers nowhere near powerful enough to make any impression on the giant seas and hurricane-force winds. With a crunching and wrenching of iron she was thrown on to rocks and listed alarmingly: the ship was barely fifty feet from the shore, but so fearsome was the weather and so hostile the sea that she might as well have been fifty miles out for all the sanctuary the distance provided. It was also so dark and the visibility so reduced by the squall that nobody had any idea of the nature of the coastline on to which they had been thrown. Captain Taylor thought the tide was on its way out and they'd be left with the hopefully simple task of walking ashore once the storm had abated. In fact the tide was coming in, the worst possible scenario for the stricken vessel.

At around six o'clock in the morning, in the village on the nearby headland, two men looking to secure the roof of a storm-battered cottage noticed the *Royal Charter*, just making out through the spray and the murk the outline of her hull in the storm and the blue dawn light. They rushed down to the rocks, barely thirty yards from the ship, but as the massive waves thundered over her and juddered against her iron hull the two men, who'd lived on and by the sea all their lives, knew there was nothing they could do.

On board, as the captain and crew debated what they should
do, a Maltese seaman who went by the anglicised name of Joe
Rogers volunteered to try to swim to the rocks with a line,
along which a bosun's chair could be passed to get people off
the ship. Like the men standing helpless on the cliffs he had
spent his life on or by the sea and knew this was the only course
of action with any chance of saving lives. He climbed out along
the jib boom to get as clear of the ship as possible, took hold of
the loose guy rope and lowered himself into the crashing waves.
Somehow Rogers managed to reach the rocks, where a human
chain of men who had gathered from the village hauled him to
safety. A makeshift bosun's chair was constructed and around
seventy people gathered at the forecastle only for a giant wave
to sweep everyone from the deck into the sea. None survived.
A few people, all men, did manage to make their way one by
one along the hazardous line, buffeted by spray and wind and
the spume exploding from the rocks below. Then, pounded by
waves and gales, with a screeching of iron that pierced the air
above the booming of the waves, the ship suddenly broke into
three pieces and the line was lost. Captain Taylor had sent the
passengers below for their own safety: most couldn't escape as
the ship's sections were torn open, flooded and rolled over in
the water.

Isaac Lewis was awaiting the bosun's chair when the ship
broke apart and he knew straight away that he was most likely
doomed. He looked into the darkness and saw the human chain
passing survivors over the rocks to safety. Among them Isaac
recognised his father. Having sailed thousands of miles from the
other side of the world, the storm had taken the *Royal Charter*
and broken her up a matter of feet from his tiny home village

of Moelfre. Standing at the sinking, tilting bow of the fractured ship, Isaac Lewis was just yards from his home and even closer to his own father. The men called to each other. In the storm it was almost impossible to hear, but the last words Isaac was heard to yell were '*O 'nhad, dwi wedi bod adra i farw*' – 'Oh my father, I have come home to die.'

Twenty-one passengers survived the wrecking of the *Royal Charter*, along with eighteen crew. All were men and none were Isaac Lewis, who, according to some reports, reached the rocks only to be pulled back by the waves and swept away to his death. It is hard to put an exact figure on the number of casualties as the most definitive passenger list went down with the ship, but the number of deaths most often cited is 459. Like hundreds of others, Isaac Lewis's body washed up on the shore the next day and he was laid out in the little church of Llangallo, a temporary mortuary and the place where he would have prayed while growing up.

Two months after the disaster Charles Dickens travelled to Anglesey to meet the local vicar, Stephen Roose Hughes, who had worked tirelessly to identify as many of the bodies as possible, as well as fielding correspondence and visits from the bereaved and the possibly bereaved in the weeks and months that followed the storm. According to Dickens, who wrote about his visit in *The Uncommercial Traveller*, Hughes wrote more than a thousand letters to the concerned and bereaved, and the sheer physical exertion of his efforts combined with the trauma of what he saw that night must have contributed to his early death little more than two years after the storm.

There was damage and suffering right across Britain that night – 133 ships are known to have been lost and the death toll reached more than eight hundred as the storm made its way from the south-west to the north-east – and this has come to be known as the *Royal Charter* storm. The scale of the death toll guaranteed that. But the *Royal Charter* disaster goes far beyond the human tragedy of that night, which is why it's so important to this story.

For me, the *Royal Charter* rounded Anglesey and sailed into a crucial turning point in our relationship with the weather. This was an era of incredible progress, and at the time the *Royal Charter* was an incredible ship. She bridged the ages of sail and steam. No longer would long journeys be dependent on a fair wind: if she hit the doldrums the engineer could stoke up the engine and her two hundred horsepower would keep her on the move, while her ironclad hull made her less vulnerable to the whims of stormy seas. It was no idle boast on the poster advertising her maiden voyage that she offered 'the only opportunity yet presented to the public of certainty in the time required for the voyage'. The weather, it was felt, no longer had the whip hand in a return voyage that circumnavigated the entire globe. There was never the 'unsinkable' claim that would haunt a later shipping disaster, but there was a quiet confidence that Mother Nature's influence was, if not negated entirely, then certainly much reduced.

In addition, the ship was almost a perfect representation of Victorian society: the rigid class divisions between the sections of the ship and the increasingly mobile nature of the Empire: reaching Australia within two months had been unthinkable a few short years earlier. The transformation of Australia from a place to send criminals to a desirable country in which to do business and even settle was enhanced by the relative ease with which one could now travel there, while the presence of so much gold and wealth on board enhanced the sense that, by hard work, speculation and determination, even the poor could make their fortunes.

Elsewhere man seemed to be taming nature. Industry and new farming methods were making the changing of the seasons less precarious to society and the economy. There were incredible advances in science and technology: earlier that year work had started on the Suez Canal, one of the great engineering projects of the age, while barely a month after the storm Darwin would publish *On the Origin of Species* and change the way we think about nature for ever. Even Charles Blondin crossing the gorge at Niagara Falls on a high wire seemed to suggest that man was besting nature.

Then came the night of 26 October 1859, when the sleek, speedy and safe *Royal Charter* was smashed to matchwood and iron shards a few miles from home after circumnavigating the globe, nearly all her passengers and crew dashed against the rocks just a few feet from safety. The weather was reminding us just who was in charge. She'd even snuck up on the ship, unleashing her most fearsome excesses before anyone really knew what was happening. It was as if the pride of Victorian maritime engineering had been deliberately picked up and thrown against the shore. There was nothing Captain Taylor could have done to prevent the fate of his ship, his crew, his passengers and himself, and he was posthumously exonerated from any blame by an inquiry. It was a reminder of man's mortality in the face of nature's whims. Even the fact that the *Royal Charter* had been smashed against such a short stretch of rocks between two beaches – running aground on either side would most likely have seen everyone saved – illustrates the malevolence of the forces at work that night.

The ship that didn't seem to want to go to sea produced two genuine heroes in Joe Rogers and Stephen Roose Hughes, but for me the *Royal Charter* storm is more about the death of Isaac Lewis. Could anything have demonstrated the merciless nature of the weather more than his being dashed to his death against the rocks on which he would have sat as a child, beneath the tiny, obscure fishing village in which he'd grown up, a matter of feet from his own father, having just sailed literally around the world?

Isaac Lewis would have known all about the weather. He'd have seen the storms, the mists, the hazes and the brilliant clear skies all around the globe. Just a few hours before the tragedy the weather had been calm, the sea smooth and he was nearly home. Had he been in Moelfre that evening he might even have spotted the signs of the approaching storm. Either way, the death of Isaac Lewis and those who perished with him that night had ramifications that went beyond a tragedy that is remembered in the area even today. For, hundreds of miles away in London, Robert FitzRoy read the reports of the dreadful scenes and tragic stories and vowed that, if he had anything to do with it, nothing like the *Royal Charter* tragedy would ever happen again.

12

DEEP DEPRESSION: THE BIRTH OF THE WEATHER FORECAST

Robert FitzRoy had predicted the *Royal Charter* storm. Not in the sense that he could have pinpointed the terrible consequences occurring on the Anglesey coast, but he did know the country was in for some unusually bad weather by the simple means of his barometer. October 1859 had been strangely warm until the night of the 19th, when in London FitzRoy's thermometer suddenly plummeted to below freezing. Yet in the far west of Britain and in Ireland the night remained warm and humid. FitzRoy knew there was a drop in pressure on the way but it didn't arrive until the 22nd, and when it came it was sudden and deep. The knowledge he'd acquired over years at sea and the more scientific approach with which he busied himself at Parliament Street told him that somewhere, probably well north of London, fierce winds and even snow were almost certainly assailing the country as he stood looking out of the window at a still London late afternoon. That evening he met visiting relatives from Yorkshire whose train had been delayed, he learned when they arrived, by snow and strong winds. Barely three days later the *Royal Charter* sailed into one of the worst storms ever recorded in the Irish Sea.

Even at a time when people might have been inured to tragedy as industrial accidents and pit disasters were almost commonplace, and ships being lost in naval battles was well within living memory, the *Royal Charter* disaster caused hand-wringing and mourning that swept the country. When FitzRoy picked up *The Times* and read of the massive loss of life so close to safety it hit him hard because this was exactly the kind of thing he was working to prevent. As he read the headlines, the reports and the names on the painfully short list of survivors he must have imagined the howling of the gales, the crashing of the waves and the cries of the drowning. In the following days, as the tales of personal tragedy and the accounts of bodies washing up on the shore kept the story alive and the nation aghast, the old sailor was racked with anguish for not somehow preventing the loss of life.

He knew such disasters could be averted for want of a few basic instruments and charts, not to mention the difficulty of persuading experienced seamen to change their ways, to erase the lore they had inherited and built up throughout their careers and instead trust a rising and falling column of quicksilver to help make their decisions for them. As he sat in his office at the Board of Trade, FitzRoy was torn between the helplessness and frustration that could easily trigger another depressive episode and his desire that such a tragedy should never be allowed to happen again. The British Association for the Advancement of Science had in 1858 put forward a proposal for the introduction of a system of telegraphed storm warnings and the loss of the *Royal Charter* confirmed for FitzRoy that new technology was the way forward. Within weeks of the storm he had managed to produce charts that showed its formation and progress with commendable accuracy, thanks to information sent in by post from his network of volunteer weather observers. If the same information could arrive with the virtual immediacy of a telegraph then it would be possible to plot and predict the paths of depressions and storms almost as they were happening.

If any good came out of the *Royal Charter* disaster it was the renewed sense of urgency it instilled even among the legendary bureaucracy of the civil service. By the autumn of 1860, less than a year after the storm, FitzRoy had in place thirteen telegraph stations around the coasts and also received reports from five key locations in continental Europe. He'd provided the equipment and instructions and the telegraph operators did the rest: every morning except Sunday FitzRoy's observers took readings of barometric pressure, wind speed, rainfall and temperature, and sent them by telegraph to London.

As part of his instigation of this telegraph network, in February 1861 FitzRoy had introduced the brilliant system of storm cones: canvas drums and triangular cones that could be suspended at different points on the coast so that shipping in the area could be warned of storms and high winds. A cone whose tip pointed upwards indicated a north wind, downward a southerly wind. A drum beneath the northerly cone or above the southerly cone indicated dangerously high winds. The genius of this system was in its simplicity: as the cones and drums looked the same from every angle they could not be misinterpreted.

FitzRoy had made incredible progress in just a few years. He'd transformed the Meteorological Department from a man in a room in a gentlemen's club to revolutionising the compilation of charts and the establishment of weather stations, making such headway that he was now able actually to predict storms at sea with impressive accuracy. Lives were being saved in numbers at which one couldn't even guess. Many lesser men would have been satisfied with this achievement but, despite being well into his fifties, FitzRoy saw this as just a springboard to further progress. He enlisted more and more people around the coasts to send him their weather readings, meaning that he could prepare ever more detailed charts. He became more confident in the term he'd chosen for his weather predictions – 'forecasts'. He'd needed a term which would indicate that what he was doing was purely scientific and nothing to do with quackery, astrology or fortune-telling. 'Prophecies or predictions they are not,' he would emphasise in his *Weather Book*. 'The term "forecast" is strictly applicable to such an opinion as is the result of a scientific combination and calculation.'

The Weather Book was no twopenny almanac based on folklore, supposition and plain guesswork; it was the first instance of the scientific application of data to predict the weather. Indeed, so plausible was Fitzroy's work that almost exactly six months after the first storm cones appeared and less than two years after the *Royal Charter* storm his first general weather forecast appeared in *The Times*, on 1 August 1861. Although the first use of the storm cones earlier in the year had been the first generally displayed prediction of bad weather to come, the forecast published in *The Times* was the first public usage of the word 'forecast' in connection with the weather, and the first publication of a general assessment of how the weather should be rather than a simple storm warning. This was the date used by the Meteorological Office to mark the 150th anniversary of the first weather forecast in 2011.

For me, though, the first use of the storm cones six months earlier was equally significant. When a clutch of telegraph operators left their stations that morning, went out to the specially constructed towers, checked their written instructions again, hoisted the cones and drums, looked up at their handiwork, looked out to sea and then back at the cones and wondered whether anyone out there would see them, the world of the weather was changed irrevocably. The *Times* forecast was undoubtedly the greater step forward in terms of wide coverage and general forecasting, but in my opinion 6 February 1861 is the day Robert FitzRoy was confirmed as a truly great man. Who knows how many lives were saved at sea that day, and on the eight further occasions the cones were deployed during that spring and summer? That, after all, was FitzRoy's main motivation: to try to prevent further *Royal Charters* and Isaac Lewises. The fact that he pioneered almost as an aside the daily ritual of newspaper and broadcast weather forecasts was an extraordinary bonus; serendipity from which we benefit today.

This was the mark of Robert FitzRoy. Employed purely to put in place a system of storm warnings for shipping, he exceeded that brief by miles. Whatever it was in his psyche that drove him to such perpetual, gargantuan effort, it was never sated. For FitzRoy the job was never done. What drove him on? The spirit of the public good played the biggest part in his motivation, undoubtedly, but his almost self-flagellatory commitment to his work must have come from elsewhere. His two most notable depressive episodes had come at times when outside factors had halted or stalled his progress: off Chile on the *Beagle*, when he had to abandon his commissioning of a schooner to survey the trickier parts of the coast when funding was refused; and in New Zealand after the news came that he was to be recalled. On both occasions he found himself either with nothing on which to focus his attention or unable to continue the job at hand through circumstances beyond his control.

In throwing himself into unhealthy levels of work was he trying to keep his mind permanently occupied? All the time he was innovating and recording, planning and plotting, was he fighting to keep at bay the personal introspection that allowed the dark fog to creep into his mind, the harpies of self-doubt and depression that had seduced him in the past? The suicides of Pringle Stokes, captain of the *Beagle*, and his uncle Viscount Castlereagh had impinged on both his personal and professional worlds: did he live in fear of the same fate if he allowed the darker side of his mind to emerge from where work kept it at bay? Were the lives he was saving designed to somehow appease the darkness and self-loathing that had always lurked in his consciousness?

Whatever the reason, FitzRoy was never satisfied. The success of the storm cones wasn't enough; they in turn led to the creation of his general weather forecasts which, while accurate more often than not, would never be precise enough for FitzRoy. Even the daily collation, composition and distribution of the weather forecasts involved enormous effort and achievement: the observation stations would record and telegraph their data to London at around eight o'clock in the morning. They would be delivered to the office on Parliament Street at around ten, and within an hour that day's forecasts for the whole of Britain and Ireland would be on their way to *The Times*, the Admiralty, Lloyd's of London and a handful of other publications. One hour. At the end of that frantic hour, once the reports had been dispatched, it was back to the job of analysing logs and past records in order to create at least an impression of the greater weather picture, with all its trends and foibles, for the benefit of those out at sea.

There was also the frustration that outside FitzRoy's immediate staff nobody seemed to match his drive, attention to detail and sense of how important the work was. This had been behind the Chilean schooner incident: to him it was obvious that a schooner much smaller than the *Beagle* was necessary to take the soundings and observations that the larger ship couldn't, otherwise the job he'd been sent to do wouldn't be done properly. Without hesitation he set about commissioning one from his own pocket because the Admiralty would be sure to reimburse him. When word came through after work was well under way that they'd refused, he was aghast at being unable to complete the job to the necessary standards for want of a few measly quid for a new boat.

Such was Robert FitzRoy's idealism in these matters that bureaucratic practicalities like budgets, not to mention the political in-fighting and petty personality clashes that blight just about every large organisation, frustrated him beyond measure. There was no better illustration than in the copy of his *Barometer and Weather Guide* of 1861 that I found in the British Library, bound in a volume of contemporary scientific publications just after a dense and lengthy treatise about the sea temperature off the coast of Scotland.

'The oldest seamen are often deceived by the look of the weather,' he wrote, 'but there is no instance on record of very bad weather such as would have involved loss of life . . . having come on without the barometer having given timely warning. By the very small expense of an establishment of barometers, so placed as to be accessible to any fisherman, boatman or others on the coasts, much loss of life as well as loss of boats and even shipping might be prevented.'

For a very small expense, much loss of life and of shipping might be prevented. This was FitzRoy's mantra and ultimate frustration during his time in the Meteorological Department. What was obvious to him were just figures on a balance sheet to the bean counters and time servers at the Board of Trade. Hadn't he proved his hypothesis in the Strait of Magellan on the journey back from New Zealand, an incident which, had he not been so insistent, could have resulted in his death and that of every person on the ship? Why could nobody else see what was so obvious to him?

In addition to his bouts of depression and frustration came the chief legacy of the *Beagle* voyage: the publication of Darwin's *On the Origin of Species*. At the time of the voyage FitzRoy had been open to new scientific theories, and indeed presented Darwin with a copy of Charles Lyell's then controversial *Principles of Geology* as a gift to welcome the young naturalist aboard. However, his first wife Mary, to whom FitzRoy had been engaged when he left and married within two months of his return (even though Darwin said that in the course of their five-year voyage he had never once mentioned being engaged), was a deeply religious woman and through her FitzRoy became convinced of the literal truth of the creation story. When Darwin's theories were published, positing that the earth and human beings were much, much older than the Bible suggested, and that we'd been descended from ape-like creatures, FitzRoy became very distressed. After all, it had been his idea to take Darwin along in the first place, and as Darwin mentioned the *Beagle* in the opening sentences of the *Origin of Species*, everyone would know that he had been Darwin's captain. Indeed, if FitzRoy is known for anything today it is for being the man who recruited and escorted Darwin on the voyage that changed the way we see the world.

At a conference of the British Association in Oxford in June 1860 FitzRoy had been invited to give a paper on the subject of British storms. He was already aware of Darwin's theories and indeed had conducted a brief correspondence with the naturalist on the letters page of *The Times*. Although he'd signed himself 'Senex', Darwin knew immediately who was behind the letters and pointed out that his correspondent had once told him that the dinosaurs had died out because they were too big to fit through the doors of Noah's ark. It was at the British Association conference that FitzRoy realised with despair just how fast Darwin's hypothesis was taking hold. His own paper to the scientific academy was part of the process that led to the introduction of the storm cones and had been prepared in reaction to the *Royal Charter* tragedy: it deserved headlines. Instead the talking points were presentations about Darwin, which bracketed FitzRoy's speech on the days either side. It is the second of these, on 'The Intellectual Development of Europe, with particular reference to Mr Darwin's work *On the*

Origin of Species', presented by John William Draper, that is still remembered today. Darwin wasn't attending in person but the formidable defender of evolution Thomas Huxley would be pitting Darwinian theories against the stoic theology of Samuel Wilberforce, Bishop of Oxford. FitzRoy addressed the debate from the floor, 'regretting' the appearance of Darwin's book and ridiculing the idea that it was a 'logical arrangement of facts', but his voice was drowned out by the massive and fervent support for Darwin, particularly after Huxley administered the *coup de grâce* following the Bishop of Oxford's question as to whether it was 'on his grandfather's or grandmother's side that the ape ancestry comes in'. Huxley's reply – that he 'should feel no shame to have risen from such an origin, but I should feel it a shame to have sprung from one who prostituted the gifts of culture and eloquence to the service of prejudice and falsehood' – brought the house down, but it would have had FitzRoy leaving the chamber a troubled man. Not only did he disagree violently with the theories sweeping the intellectual classes, he realised that he was in part responsible for their very spawning. He responded as he always did: by working even harder.

Somehow FitzRoy also found time to write *The Weather Book*, which was first published in 1862 and went on to become a bestseller. His prose may have been as dense as his mind was busy, but it remains a milestone work to this day. Subtitled *A Manual of Practical Meteorology* and weighing in at nearly five hundred pages, it's not exactly practical enough to slip into the back pocket, but is arguably the most important work on the weather since Aristotle's *Meteorologica*.

'This small work is intended for many, rather than for few,' FitzRoy writes on the opening page, redefining at a stroke the word 'small', 'with an earnest hope of its utility in daily life. The means actually requisite to enable any person of fair abilities and average education to become practically "weather-wise" are much more readily obtainable than has been often supposed.'

Later he contends that 'This book is intended to be popular – not necessarily superficial – but suited to the unpractised and to the young, rather than to the experienced and skilful, who do not need such information.'

The altruistic nature of the work is emphasised by the fact that *The Weather Book* was written outside his day job. He would write at home, late into the night, sometimes all night, before heading straight back into the office in the morning. Parts of *The Weather Book* we know now are incorrect, but most of the time FitzRoy is pretty much spot on. Reading the book one realises just how phenomenal an intellect FitzRoy had: this was a man, remember, with no scientific background beyond the ability to survey and record landscapes. Before his appointment to the Meteorological Department FitzRoy's knowledge of the weather had been gleaned in his spare time and through his experiences at sea. This man established the principles of meteorology and weather forecasting that are still in place today, while in a job he didn't undertake until he was nearly fifty years old and for which he had no scientific qualifications whatsoever.

FitzRoy's main legacy is the weather forecast itself; *The Weather Book* gives us its foundations. It is where he set down the knowledge and research he'd acquired for all to see. It's not a gripping read – FitzRoy was many things, but a prose stylist wasn't one of them – but his passion and zeal fizz out of every lengthy sentence and burst out of every perambulatory sub-clause.

By the time 1863 dawned FitzRoy should have been enjoying happy times. He sought no personal fame but his innovations were out there saving lives, and there was tangible evidence that he was benefiting the public good. Even Queen Victoria herself was consulting him before setting off across the Solent for her residence at Osborne House on the Isle of Wight, and the appearance of a royal footman at his front door became a regular occurrence. Yet, if anything, his mental state was deteriorating. People were still dying at sea for want of weather information, Darwin's theories were becoming ever more popular and his tired fifty-eight-year-old body was having trouble keeping up with the pace of his mind and ambitions. The fact that his forecasts were not accurate every time led to him being criticised in both scientific and public circles, despite the obvious benefits of his work. Even *The Times* occasionally mocked him and his forecasts in its editorials.

By 1864 FitzRoy's health was poor enough for his doctor to prescribe complete rest and the Board of Trade granted him leave to recover. He took a house in Church Road, Upper Norwood, in order to escape the noise and pace of London life, and decamped there with his second wife Maria and his family. If anything, this forced inactivity just made things worse. Left alone with his thoughts, away from the office and feeling guilty about being paid from the public purse while doing nothing, FitzRoy became restless and agitated. As the winter of 1864 became the spring of 1865 his health fluctuated wildly: some days he could barely get out of bed, on others he was defying medical advice and going into the office. Without the regular routine of work and having to spend most of his days confined in a new, unfamiliar house when sheer physical exhaustion prevented him from doing anything at all, FitzRoy's mental deterioration continued until the morning of Sunday 30 April 1865. After a restless night he got out of bed, visited his daughter Laura's room, kissed her, went into his dressing room, closed the door, opened his razor, raised his chin and passed the sharp edge across his throat. He was two months shy of his sixtieth birthday.

It is impossible to say exactly what possessed Robert FitzRoy to take his own life that morning. There was the frustration that his physical condition was preventing him carrying on the work he'd set himself, work that had no end and no limits. Over the previous year he'd become increasingly deaf, which would have concerned him greatly; it was as if he was being trapped further and further inside his own mind. His forced convalescence would have prevented him being occupied by the everyday pursuit of his boundless goals and he would have had too much time to spend with his own thoughts. The darkness that always lurked on his personal horizon was looming ever larger. As someone who was his worst self-critic, he took criticism from others particularly badly, and that only increased the more often his forecasts were published. This triggered further questioning of his abilities, opening a chink in his consciousness that would have allowed the dark thoughts to flood in. In addition, despite FitzRoy's firm belief in the literal truth of the Bible, Darwin's theories would have forced him to question the very basis of his own religious beliefs, and the fact that their foundations had been rocked by his own actions could only have increased his mental turmoil.

Whatever the internal workings of his fevered mind, on that spring morning in 1865 we lost perhaps the greatest figure in

the entire history of meteorology. The world owes Robert
FitzRoy a huge debt of gratitude. It pains me that he's still best
known as Darwin's captain – not that that shouldn't be honour
enough, even given his own personal horror of Darwin's theories – because I believe that his work in meteorology alone
should have him fêted today, more than two centuries after his
birth.

For a radio programme that would form part of the bicentenary commemoration in 2005 I was able to go inside the
house in Church Road and meet one of the residents of the flats
into which it had been divided. He was fully aware of the
house's legacy and told me that a tree believed to have been
planted by FitzRoy himself had only been cut down a couple
of years earlier. I tried to see if I could feel a sense of the man
in the house, looking out at the garden and trying to imagine
him sitting there or, more likely, pacing around, planning and
muttering and wringing his hands. Even though he had only
been there for a relatively short time this had never been a happy
house for FitzRoy: what should have been a calm, content
dotage was instead almost an enforced incarceration, a claustrophobic, increasingly silent sentence that plunged him into a
losing battle with his own demons.

The room in which he died was on an upper floor that I
couldn't visit. As I left I'd asked the man if he thought FitzRoy's
ghost was still in the house and, expecting a chuckling reply in
the negative, I saw him become thoughtful. He'd never noticed
anything, he told me, and he'd lived there for a number of years.
There had been a family in one of the upper flats a few years
earlier, however, and their young son had apparently begun
smiling and laughing while looking at an empty corner of the

room. He told his mother that there had been a man standing there wearing a blue jacket with gold bits on the shoulders but he'd gone now. When she quizzed him further he said that the man had been smiling too.

Returning to Upper Norwood six years later, little had changed. I'd noticed that the bright red door of the house had been faded to pink by the sunshine but the plaque commemorating FitzRoy's occupation was still a vivid green. I spent a few more minutes at the foot of his grave. The noise of the traffic was unrelenting here at the conjunction of two busy roads. There was nobody else around but me – at least, nobody outside a vehicle. The day was cold and grey and the odd spot of rain was in the wind. It was 6 February 2011, exactly 150 years after the first storm cones had been hoisted on Robert FitzRoy's instructions, to warn maritime traffic off the north-east coast of England there was bad weather on the way. It may not have officially been the first weather forecast but you could certainly argue that this was the 150th anniversary of the first shipping forecast.

As I began to walk back towards the railway station, somewhere a meteorologist was consulting synoptic diagrams as part of his or her preparation of the next shipping forecast, a forecast that deservedly carries the name of FitzRoy in its litany.

13

THE GREAT STORMS

When I was a small boy we lived in south-east London, in a house at the top of a hill where, late on stormy nights, the wind would tear around my window and fling rain against the glass in gusts that sounded like gravel fired from a blunderbuss. I'd pull the covers round my head and squeeze my eyes closed, fingers pushed into my ears, but even then the deep rumble of the wind would penetrate; not past my fingers but resonating through my bed and my entire body in a low, threatening boom. Sometimes an orangey flash would light up my clamped eyelids, at which I'd pull my knees up to my chest and count the seconds before the sinister *basso profundo* lurched across the sky and rumbled through my bones, each number I reached allowing me to relax just a little more until the inevitable timpani crescendo rolled around the heavens then died away.

Storms lose a little of their supernatural menace when we've grown up, yet they still induce latent apprehension or even outright fear when they break overhead. The arts use storms to invoke fear, tension and drama, perhaps most noticeably in

opera. I asked my friend Jessica Pratt, one of the world's leading operatic sopranos, about the role of storms in opera.

'Personally, when there is a storm brewing the atmospheric
pressure gives me a massive headache which makes it really hard
to sing, so I always hope that the storm will either break or pass
before I have to go on stage,' she told me, 'but in terms of storms
in opera they are often used to accompany battle, fear, murder
and betrayal. Generally they indicate the negative: summer and
winter are often used to show the passage of time. I can't tell
you how many operas I've sung in which I start out happy in
spring and in the last act go bottom-up in winter where I find
out I've been abandoned, betrayed or I just simply go mad.'

She cites some of the great operatic storms: Bernstein's
Candide, Verdi's *Rigoletto* when Gilda sacrifices herself to save
the cheating Duke, and Rossini's *Otello*, when Otello kills
Desdemona as a storm rages all around them.

'There is hardly ever an opera set on the sea or in a city by
the sea that doesn't feature a great storm,' said Jessica. The drama
of the storm suits the drama of opera as the storm is the most
dramatic weather manifestation of them all. Bad weather equals
bad things happening. We remember bad weather but good
weather isn't particularly memorable. You don't hear people
reminiscing about the Lovely Clear Night of October 1952,
for example, or pausing to remember the Mild Breeze of
St Oswald's Day. We remember those occasions when Mother
Nature loses her temper and reminds us just how easily she can
break us. Most of our worst disasters have been weather-related,
in fact. Around a million people died in the Great Irish Famine
when the potato blight was incubated perfectly by a particularly
mild and wet summer in 1845. A century earlier Ireland had

been hit by another famine that some estimates suggest killed
one third of the population, and had been caused in the main
by an exceptionally cold spell that lasted the best part of two
years, between the winter of 1739 and autumn of 1741: that
final year is still known in Ireland as *bliain an áir*, the year of the
slaughter.

The Year Without a Summer came in 1816 when a series of
global volcanic eruptions, most notably Mount Tambora in
Indonesia, caused worldwide temperatures to decrease by a
degree. A constant haze – the one that had so fascinated Luke
Howard – hung over Britain and freezing, wet weather continued for most of the year, ruining crops and leading to food
shortages and riots. An estimated sixty-five thousand more
people than usual died that year, and a hundred thousand suc-
cumbed to a related typhus epidemic in Ireland. (One positive
legacy of that year is Mary Shelley's *Frankenstein*. A Swiss holiday with her husband and Lord Byron had been ruined by the
weather, forcing them to stay indoors and entertain themselves
by telling each other stories.) The great London smog of 1952
is believed to have been responsible for an astounding twelve
thousand deaths, while even the heatwave of 2003 is believed to
have hastened the ends of more than two thousand people.

Despite these massive tolls, the most memorable bad weather
tends to come not as long drawn-out periods but in the form
of the big storms: short, spectacular outbursts of meteorological fury. Generally we are spared the hurricanes and tornadoes
that blight regions with less benevolent climates, although you
may be surprised to hear that the UK experiences more tornadoes per square mile than any nation in the world except the
Netherlands. It's just that they're not very big. One whipped

through Birmingham in 2005, causing structural damage to a church that eventually had to be demolished; another almost politely lifted off the roofs from a street of houses in north London the following year, but they are usually pretty mild: there's little danger of a girl in a gingham dress ever having cause to say, 'Toto, I've a feeling we're not in Kidderminster any more.'

While our storms may be relatively mild in global terms, their impact on our islands can be tragic and their legacies long-lasting. Three such storms in particular stand out as much for their aftermaths as their initial drama.

In early February 1287 New Romney was entitled to feel quite pleased with itself. It was a thriving port at the mouth of the River Rother on the coast of what is now East Sussex, and was doing very nicely thank you. New Romney had developed as a port along the north bank of the Rother, but over the years problems with silting led, in the mid-twelfth century, to the original town being effectively split into two as the port itself edged closer to the mouth of the river in order to keep its anchorage deep enough and the non-maritime part of the town stayed where it was. The two eventually became the inland Old Romney and the port New Romney. As one of the Cinque Ports – appointed by Royal Charter in 1155, along with Hastings, Dover, Sandwich and Hythe – New Romney enjoyed so many tax exemptions, privileges and amnesties in return for keeping military ships ready at all times that it was practically a mini-state in its own right. With five churches and several inns, the port was a successful, self-sufficient town that could perhaps have been forgiven a faint air of smugness.

Further west there was the large port of Winchelsea, with

seven hundred houses and more than fifty inns, a favoured import location for wine from Gascony and a town with a distinctly well-to-do feel. Built on a vast bank of shingle, Winchelsea was, with nearby Rye, designated an antient town: essentially a kind of second-division Cinque Port without the same level of privilege but still plenty of reason to be chuffed with itself. In previous years storms had actually breached the shingle bank a number of times and parts of the town had suffered damage from the encroaching water. The medieval chronicler Holinshed detailed a storm there: 'at Winchelsea besides other hurt that was done to bridges, mills, breaks and banks there were some three hundred houses and some churches drowned'. Matthew Paris later wrote that the storm had done 'great damage by land and still greater by sea and especially at the port of Winchelsea that is of such use to England'. So important was Winchelsea to England, in fact, that in 1281 Edward I had instructed his local steward, the terrifically named Ralph of Sandwich, to start planning a move further inland, to build a new Winchelsea.

So despite their troubles with silt and shingle respectively, in early February 1287 Winchelsea and New Romney, two prosperous port towns with bright futures, had every right to a tangible air of contentment.

February 4th dawned with low grey cloud scudding in from the north-east. The wind was gusting up, making the boats bob and knock together in the harbours, but there was nothing particularly out of the ordinary; it just looked set to be a typically blustery day on the Channel. But the wind picked up and the surface of the sea grew rougher, white tips on the waves visible as far as the horizon. It soon became clear that no one was going to put to sea that day, then barrels and carts started to be overturned as the wind grew steadily stronger. By the afternoon it had developed into a screaming, howling gale, a storm so violent it would actually change the physical geography of the country.

In mid-afternoon the sky darkened to an evening shade and the wind tore at the roofs of houses. Boats were smashed against the quays and each other and the streets were filled with bowling, bouncing debris and fallen trees, while ropes whiplashed lethally from masts. Terrified animals broke out of their pens and careered through the streets, some blown off their feet by the power of the wind, while people took shelter in churches. The sheer naked violence of this storm was like nothing anyone had ever seen before.

Contemporary accounts tell how the sea churned and boiled and turned a fiery red, leading some modern experts to wonder whether there had also been some kind of undersea eruption or earthquake. Whatever happened, whatever had caused the storm, by the time the traumatised people along the south coast unwrapped their arms from their heads and each other and gingerly ventured out from wherever they'd been sheltering, in both towns they found a very different vista from the one they remembered from a matter of hours earlier. The first thing they would have noticed was the silence. Where normally the morning air would have been filled with the sounds of a busy port town – the shouts of the traders, the bells of ships, the laughter of children – on the morning of 5 February silence covered the south coast like a blanket.

But it was what greeted their eyes that shocked people the most. The old Winchelsea was gone. Not just badly damaged; it had disappeared. There was nothing left. The shingle spit had been destroyed by the storm and the old town was now entirely beneath the sea. Houses, churches, shops, taverns, ships, people, animals: all gone. A few piles of stones would be visible at low tide for the next four or five years, but after that there would be no trace of the original Winchelsea. Today no one is sure of the exact location of the old town, so thorough a job was done by the wind and the sea.

Further up the coast, the inhabitants of New Romney emerged from whatever shelter they'd been able to find to discover an utterly mystifying transformation. A few hours earlier the sea had been washing around the ships in the harbour next to St Nicholas's Church, but now New Romney found itself a full mile inland. Tons and tons of shingle had been washed along the coast and much of it had settled in and adjacent to the town. Brand-new land stretched south to the new coastline in the distance where the traumatised people could just about make out the white foam of the breaking waves. On the morning of 5 February 1287 New Romney found itself a port town without a port. Mud, shingle and sand filled the streets to a depth of four feet or more and there was no sign whatsoever of the River Rother that had, until recently, run out to sea there. The storm had taken the river and completely diverted its course, so it now emptied into the sea at Rye, some fifteen miles away. The nearby village of Broomhill had been destroyed, and at Hastings the storm had torn away a large section of the cliffs, taking part of Hastings Castle with it and blocking most of the harbour.

The legacy of that night can still be seen in New Romney more than seven hundred years later. If you walk down Church Approach from the high street, just after the tourist office you come to St Nicholas's Church. It is a large Norman structure that looks out of place in what's now a small town, but that's not what makes it unusual. As you approach the main door of the church you find you need to walk down four steps to reach it. Being by far the sturdiest building in the town, this church was the only structure to survive the storm. When New Romney was inundated by silt, sand and shingle, what became the new street level was four feet higher: the church floor is all that remains of the original level of New Romney, and is now effectively underground. Inside the church, look closely at the pillars that line the nave: each one has an orangey-brown discolouration to a height of around four or five feet: they're still stained by the muddy water of more than seven centuries ago. Walking past the church you'll see the road leads in a straight line towards the horizon about a mile away. The sea is just behind it. It's extraordinary to think that the smart, modern maisonettes on the left, just beyond the churchyard wall, the ones with the hatchbacks in the drive and the manicured patches of lawn, are on the exact spot where New Romney harbour once stood. One wonders whether the dreams of the middle managers and estate agents living in them are ever disturbed by the sounds of crashing waves and the wind screaming through the rigging of ancient ships.

From the thirteenth century we move forward half a millennium to the storm that lashed southern Britain on the night of 26 to 27 November 1703, a tempest that has become the yardstick by which all our weather events are measured. The 1987 storm was bad enough for those who lived through it, but that wind was just a bit lively when compared to what was unleashed across the same area 284 years earlier. The level of devastation was unprecedented: thousands of people lost their lives and count-less buildings were destroyed, while one fifth of the English Navy was wiped out.

By 1703 a scattering of people had begun to keep weather records. There were even a few with barometers, meaning com-parisons could be made across wide geographical areas even if a reliable thermometer was still a decade or more away. In addi-tion, the 1703 storm produced what many argue is the first piece of real journalism: Daniel Defoe's record, published in 1704 as *The Storm*, is the principal reason why we know so much about the important events of that fateful night.

The weather had been turbulent for most of the previous week, but had provided absolutely no hint that the largest storm in a millennium or more was on its way. Strong winds had been blowing in from the south-west for several days, but on 26 November they became particularly fierce. A tornado was reported in, of all places, Oxford as a deep depression – one barometer in Nottingham recorded a low of 950 millibars, an extraordinarily low figure only usually found in the eye of trop-ical storms – moved east-north-east, triggering a wall of wind roughly three hundred miles across that passed over southern Britain and the northern fringe of the European continent, steamrollering everything in its path. By evening the wind was estimated to have increased to speeds as high as 170mph: when you consider that on the Beaufort scale, still a century away, hurricane force 12 begins at 70mph, the terrifying nature of this storm is obvious. The scale of the devastation was unprece-dented, even to a population to whom the Great Plague and the Great Fire of London were still relatively recent memories. As

Defoe put it regarding the latter, 'that Desolation was confin'd to a small space, the loss fell on the wealthiest part of the People; but this loss is Universal and its extent general, not a House, not a Family that had anything to lose, but have lost something by this Storm'.

Defoe estimated that more than eight thousand lives were lost that night, but fewer than two hundred of them were on land – again it took the sea to prove our vulnerability to the weather. The seamen killed that night were mostly from naval ships, hence that incredible figure of one fifth of the English Navy perishing in the space of a few hours, despite being nowhere near a battle. Who knows, maybe it's in the echoes of lives lost like these that the resonance of the shipping forecast lies? Even in its benevolent poetry it is a sobering reminder of what a fearsome duo the sea and the weather can make when you're clinging to a vulnerable, low-lying bunch of little islands.

As many as two hundred ships were destroyed that night, with some estimates putting the total death toll at sea as high as ten thousand. Many ships were sheltering at Goodwin Sands off the Kent coast and were wrecked there, while others were scattered to all parts. The main naval fleet had just returned from the War of the Spanish Succession under Sir Cloudesley Shovell: these were battle-hardened men fighting in vain for their lives as the wind and the sea smashed their ships to pieces. Famous fighting ships like the *Stirling Castle*, *Mary*, *Northumberland* and *Restoration* were sunk, and while twenty-one sailors survived the *Stirling Castle* the others went down with all hands. There was the extraordinary story of a seaman called Thomas Atkins: the only survivor from the *Mary*, he was washed onto the deck of the *Stirling Castle*, then as she sank he was washed to safety in

her one surviving lifeboat. Cloudesley Shovell and his ship, the
Association, were missing for several days and prayers were
offered in the press. People feared the worst, but it turned out
that Shovell had ordered the masts to be cut down, and the
helpless ship was carried all the way to Gothenburg, returning
weeks later.

Inland, a storm surge along the Severn Estuary flooded Bristol
and a number of nearby coastal villages, accounting for eighty of
the deaths on land. At Wells Palace Richard Kidder, the Bishop
of Bath and Wells, was killed alongside his wife when a chimney
collapsed in on top of their bed. Perhaps unsurprisingly, some saw
this as an act of divine disapproval: Kidder was ill-liked, having
replaced Thomas Ken, who had been removed from his post,
despite being hugely popular among High Church followers, for
refusing to swear the oath of allegiance to William of Orange.
After the storm a poem appeared, supposing that 'Yet strictly
pious Ken! /Had'st Thou been there, /This Fate, we think, had
not become thy share; /Nor had that awful Fabrick bow'd'.

Of the non-human casualties, the best-known was Eddystone
Lighthouse, which was swept away with its staff and its architect
Henry Winstanley, having only stood on its rock nine miles off
the coast of Cornwall for five years. Winstanley had just happened to be there that night to oversee some adjustments.
When, days after the storm, a ship finally reached the rocks all
that was found were a few remains of the lighthouse's foundations. According to legend, on the night of the disaster a scale
model of the lighthouse in Winstanley's Essex home fell to
pieces at the height of the storm.

Across the country some eight hundred houses were
destroyed. A thousand barns and outbuildings were flattened in

Kent alone, a hundred churches – including Westminster Abbey – lost their roofs and more than a million trees were uprooted. Fifteen thousand sheep were drowned in the flooding and four hundred windmills destroyed, either blown down altogether or their sails set spinning so fast their axles overheated enough to catch fire and burn the mills to the ground. Another tornado was reported, this time at Whitstable, one strong enough to dump a ship eighty yards inland and leave a cow stuck in the higher branches of a tree. The damage came to an estimated cost of six million pounds, of which one third was accounted for in London. After the Great Fire, legislation had been passed banning thatched roofs in the city: on the morning of 27 November the streets were covered with broken tiles. The price of tiles immediately shot up by 400 per cent, which should have been good news for Daniel Defoe, but wasn't. But then he was having a pretty rotten time of it anyway.

Until that night the future author of *Robinson Crusoe* and *Moll Flanders* would not have regarded 1703 as a particularly vintage year: half of it had been spent in hiding and the other half in prison. He'd anonymously published a satirical pamphlet called *The Shortest Way with the Dissenters* in which he posed as a hysterical High Church Tory minister proffering the opinion that religious dissenters in Britain be treated in the same way that the Huguenots had been dispensed with by Louis XIV in France. The pamphlet caused such outrage that Queen Anne herself called for the author to be brought to justice and a reward of fifty pounds was put up for information about the whereabouts of one 'Daniel de Fooe'. Hence Defoe spent the first five months of 1703 in hiding, possibly even in the Netherlands. At the end of May he was captured at a house in Spitalfields – iron-ically one belonging to a Huguenot weaver – and appeared in court in July, charged with seditious libel. He was fined two hundred marks, ordered to spend three days in the pillory and was then detained in Newgate Prison at Her Majesty's pleasure. He was also ordered to stay out of trouble for seven years, and by 'trouble' the authorities meant that he shouldn't publish any-thing at all in that time. Being Defoe, before the year was out he would produce at least twenty works, including the poem 'Hymn to the Pillory', which was published on the very first day he appeared in the stocks where, as it turned out, nothing more unpleasant than flowers was thrown at him for the entire three days. The four months he went on to spend in Newgate would not have been quite as pleasant as having carnations glance off his forehead, but as luck would have it he was released in early November, three weeks before the storm that would change his life and fortunes entirely. In the short term, however, he was in fact missing out on a fortune. By 1696 Defoe had opened a brick works at Tilbury in order to pay off debts that ran to an astonishing seventeen thousand pounds. It was initially a successful venture – the bricks that built the Royal Naval College in Greenwich came from Defoe's yard – but, having been neglected while Defoe was in hiding and in prison, the business fell into disarray and had gone bust barely weeks before the storm. However, his missed business opportunity would be compensated by his grasping a great literary opportunity.

In the days following the storm, as the country picked its way gingerly through the wreckage and hundreds of bodies began to wash up on the shore, it became clear that this had been a cataclysmic event. It was 'the greatest, the longest in duration, the widest in extent of all the Tempests and Storms that History gives an Account of since the Beginning of Time', according to Defoe. On 12 December Queen Anne, who from her bedroom had watched the trees being ripped out of the ground in St James's Park before being whisked to an underground room for her own safety, declared a day of public humiliation for 19 January, when prayers would be offered in an attempt to appease God's wrath against the 'crying sins of the nation'. The clergy took the opportunity to unleash some Old Testament fire and brimstone over what was, to all intents and purposes, a distinctly Old Testament event: what greater expression of God's wrath could there be than a tempest that laid waste to huge swathes of the country and the cream of her maritime protectors? Indeed, so significant was the storm in the popular consciousness and religious climate of the times that sermons were still being delivered on the subject more than eighty years later.

In many ways, the storm occurred at a tipping point in our history. While the majority still regarded it as an act of God, the fledgling scientific approach spearheaded by the likes of Descartes and Torricelli meant that the events of that November night represent a gateway between the ancient and the modern. It also came at a time when the press was in its infancy. The *Daily Courant* had only nine months earlier begun publication as the first regular English-language daily newspaper in the Western world. It joined a stable of less regular publications, all of which really liked the word 'post', such as the *Post Boy*, *Post Man* and *Flying Post*, all of which had appeared when licensing of the press ended in 1695. Weekly journals were also appearing, for example the *Observator*, which had opened the previous year.

The storm was the perfect story for these publications and for the first time the discussion of major events could go on record across the nation, but it is Daniel Defoe to whom we owe the most for our knowledge of the storm. He produced three works related to the disaster: *The Layman's Sermon*, a pamphlet published in February 1704; a set of satiric verses called *The Storm: An Essay*, which attacked the High Church view that this was a storm sent on their behalf ('They say this was a high church storm, Sent out the nation to reform; But th' emblem left the moral in the lurch, For 't blew the steeple down upon the church'); and finally the work that is still in print today, more than three hundred years on: *The Storm: or A Collection of the Most Remarkable Casualties and Disasters which Happen'd in the late Dreadful Tempest, both by Sea and Land.*

It's a terrific, landmark work: Defoe had sensed immediately that this was an important moment in history and needed to be recorded as such. He could have turned it into a novel, he could have written another political or religious polemic, he could have done nothing at all. But instead he used the new outlet of the press to gather stories and accounts and set about turning them into as scrupulously accurate a record of the events as he could. Defoe placed advertisements requesting information in the *Daily Courant* and the *London Gazette* in the days after the storm and wrote to a number of people around the country asking them for their accounts of the night. 'Preaching of sermons is speaking to a few of mankind,' he wrote as the opening words of his preface, 'printing of books is speaking to the whole world.'

Defoe didn't just go for the sensational cow-up-a-tree stories either. For example, he prints a letter from William

Derham, a clergyman from Upminster in Essex and one of the most important weather recorders of the age. Derham had taken barometric readings during the storm and compared them with those taken by his friend Charles Towneley in Burnley, Lancashire. Although their readings were not taken at exactly the same times and were so far apart geographically, it's not stopped meteorologists being able to make informed assumptions about the scale and nature of the storm purely on the basis of their readings. Defoe collected more than fifty first-hand reports from a wide area as well as writing up details from many more locations himself. He even gives accounts of damage in northern France, and quotes observations made in Delft by the microscope pioneer Anton van Leeuwenhoek. It is incredibly rare for an event of any kind from this period to be recorded in one volume but in so many voices. Most historical accounts are second- or third-hand; in few of them can you actually hear a voice coming down the centuries. When you read *The Storm*, however, you are listening to the voices of real people articulating their first-hand experiences of a momentous event. In most cases this is the only occasion these people emerge from the fog of history: clergymen, clerks, book-keepers, soldiers, all of whom decided to sit down and record their experiences of the greatest storm in our history.

As well as giving us the accounts of dozens of people Defoe relates his own experiences too. He tells of how two days earlier he had 'narrowly escaped the mischief of part of a house', a lump of masonry or a chimney brought down by earlier high winds which, had it scored a direct hit, would have deprived us of *Robinson Crusoe*, umbrella and all. On the night of the storm itself he was in a house on the outskirts of

London; we're not sure exactly where, but possibly Islington. That morning he had noticed the needle on the weather glass in the house had sunk lower than he had ever seen, so low in fact that he even wondered whether the children of the house had been messing about with it. He also gives us his account of the following days, as the shattered south of the country tried to come to terms with what had happened. 'The streets lay so covered with Tiles and Slates, from the Tops of the Houses, especially in the Out-parts, that the quantity is incredible, and the Houses were so universally stript, that all the Tiles in Fifty Miles round would be able to repair but a small Part of it,' he wrote, presumably through vision blurred by tears of misfortune at the thought of his recently bankrupted brickworks. He walked down to the Thames and estimated that some seven hundred ships had been thrown together by the wind and the churning river there alone, some badly damaged, most destroyed altogether, and he noted the damage to some of London's most famous buildings, such as St James's Palace, from whose windows Queen Anne had been escorted at the height of the storm.

The Storm is not one of Defoe's best known works, but it's an important one in a number of ways. First, it's a vital piece of history, a primary source for one of our most extraordinary weather events. Second, it's a pioneering work of journalism way ahead of its time. And third, it captures a major natural event at a time when the balance between religion and science was at its most delicate. That the greatest storm ever should occur just when our study of the weather was beginning to progress beyond God and Aristotle and into the realm of science is coincidence enough; that one of our greatest writers was there observing ensured that its tragedy and import were not left in the fog of history.

Two hundred and fifty years later there were yet more men out on the sea at the mercy of the weather.

As night fell on 31 January 1953 the crew of the Fleetwood trawler *Michael Griffith*, some way off the south tip of Barra Head in the Western Isles of Scotland, were preparing for their second night away from the Lancashire port. She'd had to put back in not long after departing to repair a faulty pump, so the thirteen men on board would have been keen to make up time. They had not seen another ship since passing fellow Fleetwood fishing vessel the *Aigret* the previous night, but the crew, with the possible exception of sixteen-year-old novice deckhand George Parlin, was well used to being cut off from the rest of the world. What none of them could have known was that they were about to become the first victims of a confluence of weather events that would cause mayhem around the coast of Britain, as part of the worst natural disaster of the twentieth century.

The last contact from the *Michael Griffith* was an urgent radio message – 'Full of water, no steam, helpless' – in the early hours of 1 February, and despite a prolonged air and sea search no trace was ever found of her or her crew beyond two lifebuoys that washed up on the coast of Northern Ireland. Outside of Fleetwood, the loss of the *Michael Griffith* would be largely forgotten in the wider picture of a vicious, epoch-defining storm.

It emerged from an intense, fast-moving centre of low pressure that had formed off the Azores and travelled up towards the west coast of Scotland, where the poor unsuspecting *Michael Griffith* awaited. Later that day Britain's first purpose-built roll-on/roll-off ferry the *Princess Victoria* was hit by the extraordinary winds on her way from Stranraer to Larne, took in water through her stern doors in the steepling seas and sank with the loss of 133 lives, including the deputy prime minister of Northern Ireland and the MP for North Down (in an echo of the *Royal Charter*, she was a pioneering vessel that went down in the Irish Sea leaving only a handful of survivors, all of them male).

The depression, influenced by a strong anticyclone to the west of Ireland, moved east just below the Faeroe Islands, when a then-record wind gust of 125mph was recorded in the Orkney Islands, and then headed south-east down through the North Sea. As the storm system continued to move south through Saturday the 31st, high pressure continued to build over the eastern North Atlantic, increasing the pressure gradient over the North Sea and squeezing the isobars of the storm ever tighter. The strong winds pushed Atlantic water into the North Sea, creating a massive sea surge that forced the sea south into the shallow bottleneck between East Anglia and the Netherlands, just as high tide was approaching and two days after a full moon had helped to create higher than usual spring tides. The surge, the biggest ever recorded in Britain at over eight feet above the normal sea level, overwhelmed defences in Britain and Holland, flooding more than 150,000 acres of land in England and claiming 307 lives, while in the Netherlands, and Zeeland in particular, more than 1800 people were drowned. More trees than in the average annual timber harvest were felled in north-east Scotland as the wind tore trees out of the ground in its rampage around the core of the depression.

The UK government inquiry that followed the storm

revealed some frightening statistics: as well as the 307 deaths
there had been around 1200 breaches of flood defences along
the east coast, more than 30,000 people had had to be evacuated
and 24,000 houses were damaged, of which five hundred were
completely destroyed. The floodwaters had encroached more
than two miles into England. Cleethorpes was totally inundated
and not a single house was left undamaged. Raging seas crashed
through the banks of the Wash, killing sixty-six people in the
village of Heacham and fifteen in nearby King's Lynn. Thirty-
eight people died at Felixstowe when their prefabricated houses
were swept away, while the village of Jaywick in Essex lost
thirty-seven people. Low-lying Canvey Island was devastated,
with the loss of fifty-eight lives, while more than a thousand
homes were flooded in east London as the storm water surged
up the Thames as far as Chelsea Embankment.

For all the damage and loss along the east coast, the most visible
legacy of the 1953 surge is the Thames Barrier, the second-
largest movable flood barrier in the world after the Eastern
Scheldt Barrier in the Netherlands. Better systems of communication were put in place – the death toll could have been much
reduced had the word been able to spread ahead of the storm
but, with few people having telephones and what lines there
were damaged by the winds anyway, in most places the surge
arrived completely unannounced.

The Great North Sea Surge of 1953 has, despite being one
of the worst natural disasters in our recent history, been strangely
forgotten outside the areas affected on the night. There is some
grainy, evocatively silent newsreel footage on the internet if you
look hard enough, but the victims of the storm, from the crew
of the *Michael Griffith* to the people of Canvey Island drowned

in their homes, deserve better. A rare, almost unique coincidence of factors it may have been, but Britain's flood defences and response systems were shown to be hopelessly inadequate. The worst affected were, of course, the poor, such as the entire families drowned in the debris of their flimsy wooden Felixstowe prefabs. The defences are better now, but we would do well to remember that the weather is no respecter of class or money.

14

'I MERELY ATTRACT THE CLOUDS AND THEY DO THE REST': WEATHER CRACKPOTS, CHARLATANS AND DREAMERS

There are many weather heroes in this story. The pioneers, the thinkers, the scientists, the enthusiasts who have all helped to create the weather forecasts and understanding we have today. But the highway of weather progress has verges crowded with those who ran off the road: the fakers, the mountebanks and meteorological find-the-lady merchants; the eccentrics, the cranks and the well-meaning, intelligent thinkers who just got it wrong. As mankind began to explain the previously unexplainable in most walks of life, the weather proved to be one of the most stubborn bastions of mystery. Its uncertainty and arbitrariness made it intriguing and mystifying: while the likes of Luke Howard and Francis Beaufort could categorise the weather, it was one of the few aspects of nature that couldn't be tamed and made to serve. It is no wonder, then, that the weather has attracted its fair share of crackpots, charlatans, dreamers and the just plain wrong. Something as hard to define as the causes and behaviour of weather led to some extraordinary theories finding rational traction, no matter how bananas they might seem today.

Albert Stiger probably falls into the latter category. A skinny, bearded, intense-looking man with sunken cheeks and a jutting chin, Stiger was in the closing years of the nineteenth century the mayor of what was then the Austrian town of Windisch-Feistritz (it's now in eastern Slovenia and known as Slovenska Bistrica). As well as sounding agreeably like a sneeze, Windisch-Feistritz was at the heart of Austrian wine country and Stiger himself was a grape producer. However, a series of fierce hailstorms during the 1890s had badly damaged several harvests across the region and, as mayor and someone whose livelihood depended upon the grape harvest, Stiger watched the stinging volleys of frozen droplets battering his carefully tended vines and realised he had to do something.

There had always been a widely held belief in the region – and beyond – that loud noises could dissipate storms. As long as there had been gunpowder, Austrian farmers had been taking potshots at the sky whenever a cumulonimbus hove into view. Church bells were rung before and during storms, and bells were even produced with words from Friedrich Schiller's *Song of the Bell* engraved on them: *vivos voco, mortuos plango, fulgura frango* ('I call the living, I mourn the dead, I break the lightning'). So many bell ringers were being killed by lightning striking the bells and fizzing down the wet ropes that in 1750 Archduchess Maria Theresa outlawed the practice.

These folk memories stirred somewhere in the back of

Stiger's mind and he came up with the brainwave of constructing a series of massive cannons to fire at the sky whenever the air suddenly calmed, which he took as a sign of impending hail. This wasn't just bowing to superstition, there was a scientific rationale of sorts: Stiger believed that if he fired particles of shot into the clouds they would attract condensation and fall to earth as harmless rain before they could freeze and shower the vineyards with hail.

In the early summer of 1896 he placed half a dozen cannons made from a series of mortar-like tubes on the hillside and waited. Sure enough, on 4 June he sensed the temperature drop and the air grow still, an unmistakable sign of a potential hailstorm. On his signal the mortars let go with a thunderous volley of artillery that echoed around the hills. The clouds still massed overhead, their bases tinged with the dirty brown of impending release. Stiger and the townspeople looked anxiously at the sky and waited.

Within a minute or two they felt the first drops on their upturned faces before the ground hissed and the leaves flickered – with rain. People rushed to Stiger with their congratulations, which the mayor acknowledged with a grim smile, stopping short of joining in the general whooping and hollering, although having seen his stern-looking photograph I'd say that Stiger hadn't done much in the way of whooping and hollering in his lifetime anyway. At this early stage of the cannon experiment he was hopeful that his idea had worked, but couldn't be sure. Then over the next few hours news arrived that other towns and villages in the region were reporting hailstorms: Windisch-Feistritz seemed to be the only place to have had good, old-fashioned, harmless rain.

Feeling vindicated, Stiger took to battering the skies with heavy artillery with terrific enthusiasm. That summer alone saw the cannons unleashed on forty occasions, and the town and its environs experienced no hail at all for the entire season. Stiger began to expand his operation and his half-dozen mortars soon developed into huge metal cones thirty feet high, widening from a narrow shaft at the base to a gaping blunderbuss of an opening at the top, looking for all the world like a row of giant gramophone horns. Word of Stiger's success spread, to the extent that the Austrian government sent him consignments of disused funnels from old railway engines free of charge. Within a year the local hillsides were dotted with specially constructed cabins half a mile apart, from whose roofs extended the barrels of what were becoming known as hail cannons or Stiger guns.

For the best part of a decade the hail-cannon craze swept across continental Europe in a deafening wave of exploding ordnance and startled birdlife. In 1899 Italy was boasting of two thousand hail cannons unleashing merry hell at the heavens, and by the time the calendar had flipped over into the twentieth century there were in excess of ten thousand hail cannons in more than twenty countries. Insurance companies offered lower premiums to wine producers close to the guns and there was even a 'hail prevention expo' in Lyon in 1901, at which cannon enthusiasts could ooh and ahh at a range of designs.

Meanwhile, back in Windisch-Feistritz, the spiritual home of anti-precipitation artillery, two hundred cannons were pointing at the heavens by 1902, with the same number reported just across the Italian border in Castelfranco-Veneto.

Which is when things started to unravel.

Just as it started to seem as if Europe had actually declared war on the sky, not to mention the number of injuries and even deaths caused by the old principle of what goes up must come down, people started actually to examine the results of the massive Stiger cannonade, an exercise that revealed no indication that the hail cannons were having any effect whatsoever on the prevention of hailstorms. In fact, the Windisch-Feistritz district had one of the *worst* records for hailstorms anywhere in central Europe, despite the huge quantities of ordnance being twanged at the unsuspecting clouds. From 1902 to 1904 there was a series of severe hailstorms in both Windisch-Feistritz and Castelfranco-Veneto, which meant that within a year or two the craze for hail cannons had died out nearly everywhere. The desire to believe that man had at last found a way of controlling at least one aspect of the weather had combined with the happenstance of two or three years of fewer hailstorms to create an illusion of victory. Add in the opportunity for men to build big guns that made a hell of a lot of noise and it's no wonder the enthusiasm for hail cannons swept across the continent, until a few spoilsport scientists pointed out that the whole project made not the slightest difference.

Stiger died not long afterwards and it appeared that the idea went with him. Every now and again the hail cannons make a comeback. There was a spate of them in the US during the 1950s and 1960s, and even today a few farmers there deploy them. They are apparently cheaper than taking out hail insurance. It's not just farmers, though: in 2005 a Nissan car plant in Mississippi was the target of local residents' ire when they started firing off a hail cannon every six seconds whenever there looked to be the faintest possibility of a hailstorm brewing somewhere in the vicinity. I suppose if I had twelve thousand brand-new cars with immaculate paintwork sitting out in the open I'd probably take any precaution going too, but every six seconds? It must have been like living next door to the Somme.

When even a major multinational company starts firing guns at the heavens it proves that we really do want to believe we can control – or at least influence – the weather. No wonder that when someone like Albert Stiger comes along and professes to have an answer as to how we can influence the weather for the betterment of mankind, people want to believe him. The history of our relationship with the weather is packed with examples of a willingness to believe that Mother Nature can be brought to heel. In some cases the belief is so strong it can propel a well-meaning propagator of mistaken methods – or a total charlatan – to a considerable level of notoriety, even to the point of having an entire winter named after them.

This is exactly what happened to Patrick Murphy. Originally from Cork, he moved to London in 1822 to escape poverty and famine and by the 1830s was writing books of a scientific bent on subjects such as gravity, magnetism and the solar system. Many other people were doing likewise and Murphy's efforts, containing nothing significantly groundbreaking, would have been forgotten with the rest of them were it not for the events of one day in 1838.

In 1837 Murphy had published his *Weather Almanac on Scientific Principles Showing the State of the Weather for Every Day of the Year 1838*. Like most such texts Murphy's predictions were based purely on supposition, with no scientific basis beyond his own unsubstantiated theories. He believed the weather could be predicted by methods such as an examination of the moon and a close study of animal behaviour. Essentially, his methodology came down to what the man in the moon was telling him and whether the cows were lying down. Not surprisingly, his forecasts for just about every day of the year were completely wrong. With one exception. For 20 January 1838 Murphy had predicted: 'Fair, and probably the lowest degree of winter temperature'. That day turned out to be the coldest day ever recorded in England, with temperatures dropping as low as $-20°$ Celsius in London. The Thames froze; people could cross the river on foot as far west as Hammersmith and as far east as Tower Bridge. The ice was so thick at Hammersmith that people could walk out on to it and build fires – a whole sheep was roasted mid-river. It led to an extensive period of cold for Great Britain, and an extensive period of raking in piles of cash for Mr Murphy. Once word got out that he had predicted the coldest day on record – even though he'd done so in the vaguest of terms – booksellers were inundated with requests for copies of his work and his publisher's office was almost ransacked by a crowd of frantic almanac seekers. Murphy's booklet ran to an extraordinary forty-five editions and was believed to have made him a sum in the region of three thousand pounds, an absolute fortune in those days. So famous did his prediction become, in fact, that the winter of 1838 would be remembered as Murphy's Winter.

Unfortunately for Murphy, he couldn't repeat his success. Future almanacs – he produced them up to 1845 – were wildly inaccurate and sold poorly, while he lost his fortune speculating on the corn market. A correspondent in *The Times* summed up Murphy's forecasts in verse:

When Murphy says frost, then it will snow.
The wind's fast asleep when he tells us 'twill blow.
For his rain we get sunshine, for high we have low,
Yet he swears he's infallible — weather or no!

Patrick Murphy died in 1847 and has been all but forgotten, even though his fame at the time was phenomenal. There was even a play about him: *Murphy's Weather Almanac: A Farce in One Act* by William Rogers, which was performed at the Sadler's Wells theatre in 1838.

It may seem strange, given that we spend most of our time hoping it doesn't rain, but most of the kooks, hornswogglers and humbug merchants in the history of weather have been concerned with actively encouraging the stuff. As we have seen, however, rain forms the very basis of life on earth and our fragile existence depends on its regular appearance. When it doesn't appear crops are threatened and the ecosystem is endangered. People become jumpy, then desperate. And when people are desperate they're far more prepared to believe someone with a winning smile and confident patter. This leads us into the story of the rainmakers, a story we'll begin in the world of animation.

It's not often Porky Pig is cited in the history of weather, but bear with me. One of Porky's first starring vehicles was *Porky the Rainmaker*, which was produced in 1936. As the cartoon opens Porky and his father are surveying the effects of an extreme drought on their farm in the American Midwest. The crops are dying, fruit is withering on trees and such is the heat that when we're shown a patch of eggplants we see each one burst open in turn to reveal a fried egg inside. The chickens have gone on strike, marching around the yard under a banner declaring 'No feed, no eggs'. 'Worry ... worry ...' declares Porky's father, leaning on a fence with his head in his hands. He reaches into his overalls, produces the family's last dollar and sends Porky into town to buy feed for the animals. When Porky arrives he is about to go into the feed store when he's distracted by a meteorological medicine show: there's a pig wearing a derby hat and an impressive moustache standing on a wooden dais addressing a small crowd about the benefits of his rain pills, which he claims can make it rain on demand. By way of demonstration he hands out umbrellas to the crowd, takes out a blow pipe and *phuts* a rain pill into the sky. It explodes high in the air, the sky immediately darkens and it begins to pour with rain. An astounded Porky hands over the last dollar and is given a box of pills, not just to encourage rain but also effective for wind, cyclone, snow, ice, lightning, thunder and sun. Understandably, Pa Pig isn't impressed with Porky's purchase and knocks the box from his hand, scattering the pills across the yard. The animals eat most of them – with hilarious consequences – until Porky wrestles the goose for a rain pill, forcing the bird to spit it into the air. The pill explodes, the skies darken and it begins to lash with rain. Immediately the corn comes back to life, the fruit plumps up on the trees, the henhouse is raised into the sky on a pile of freshly laid eggs and Porky and his father break into smiles.

While *Porky the Rainmaker* owes much to *Jack and the Beanstalk*, there was method in the Looney Tunes madness. The consequences of drought on a large scale were much more serious in the US and pluviculture – as the encouragement of rain is known – had the potential to be big business in the early twentieth century. All sorts of snake-oil salesmen and wackos began flogging miracle rain providers, with some of them becoming famous the world over. Take Robert St George Dyrenforth, for example. Like Stiger he believed that making loud bangs in the air could affect the weather, but in his case he thought it could produce rain from a clear blue sky. Born in Chicago in 1844 he claimed to have been educated in Germany, earning a doctorate in engineering from Heidelberg University, but like many of Dyrenforth's assertions this proved tricky to verify. He also said he had been a war correspondent during the Austro-Prussian War of 1861 and after returning to America became a major in the Union army. Once the Civil War had ended he qualified as a lawyer and worked at the US Patent Office, but it was rainmaking that appeared to be his true calling. He subscribed to the centuries-old view that war caused rain to fall because the noise and heat of battle rose to disrupt the atmosphere enough for it to discharge water. An intelligent and widely read man, Dyrenforth believed that when there was an explosion in the clouds there also occurred 'something in the nature of a vortex, a momentary cavern into which the condensed moisture is drawn from afar, after which the explosion may squeeze the water out of the air like a sponge'.

What he proposed was a *Wacky Races* of ordnance heading into the sky: balloons, dynamite, kites, smoke bombs, mortars, even fireworks. When a series of droughts in 1890 threatened the harvest across wide areas of the Midwest, Dyrenforth persuaded the US Weather Bureau to petition the government with the object of giving the go-ahead for investigations into methods of creating rain. Dyrenforth, naturally, put himself forward to lead the research. Awarded the funding he'd asked for, Dyrenforth hared into the project with the zeal of a man who could at last put into practice things he had only previously been able to ponder.

In attaching a scientific basis to his theories Dyrenforth maintained that if you sent up a balloon filled with two parts hydrogen and one part oxygen timed to detonate inside a cloud, the additional moisture released into the atmosphere would be enough to wring a steady rainfall from the previously arid heavens. Given the relatively tiny size of any balloon compared to the wide open spaces of the atmosphere, he would probably have had the same impact by flying over the cloud in a plane and tipping a glass of water out of the window, but the rain-maker wouldn't be dissuaded and arranged a demonstration. First of all, he gathered a crowd of dignitaries at his country estate outside Washington and sent up a balloon ten feet across that carried a basket packed with mining explosives. At the appointed height the explosives detonated with a mighty crack that echoed around the hills for several seconds. The crowd waited. And waited. But all that fell to earth were a few shreds of smouldering balloon, while the only tangible thing the experiment produced was a strong letter of complaint from his neighbour, quite rightly up in arms about having his afternoon disrupted by an almighty explosion that rattled the windows of his farmhouse and caused his cattle to stampede.

Nevertheless, Dyrenforth talked a good game and, given the desperate straits being endured by frightened and desperate farmers, people wanted him to succeed. A rancher on the High Plains of Texas invited Dyrenforth to conduct an experiment on his land, and such was the excitement at this apparent prospect of man controlling the skies that the railroad even offered to transport him, his accomplices and his equipment there for free.

It was now August 1891. That part of Texas had been in its summer dry spell but the rainy season was due to commence. A charismatic showman, Dyrenforth gave a dramatic speech while clad in cavalry boots and his trademark pith helmet, announcing how his revolutionary methods would bring moisture to the plains pretty much on demand. The following day, as his staff of more than twenty set about preparing the explosives ready for a formidable barrage at the skies, it began to rain. So desperate were people to believe in Dyrenforth that even though the US Weather Bureau had already forecast rain for that day they spoke in excited whispers that he'd brought water from the skies simply by being there and preparing his equipment.

On 17 August everything was finally ready and in the late afternoon Dyrenforth gave the signal to unleash the ordnance. Balloons rose, mortars were fired, fireworks whooshed into the sky. The smoke cleared, but there was no rain. Evening came, there was no rain. Night fell and there was still no rain. Dyrenforth paced back and forth in the street, staring up at the clear, star-speckled sky. By dawn the clouds still hadn't gathered and as morning became lunchtime it appeared that Dyrenforth's weather artillery had failed. In the late afternoon, almost exactly a day after the sound of the broadside had died away to nothing, there was a light and brief shower nearby. Dyrenforth, of course, claimed victory.

A week later he ordered yet more explosives and equipment and set off a merciless barrage that carried on for almost the whole day, not subsiding until eleven o'clock that night. At three o'clock the following morning Dyrenforth heard distant thunder to the north but it came no closer. Again he claimed the storm had occurred directly as a result of his methods: hadn't the breeze been a southerly one, carrying the fallout from his work in the very direction of the storm?

He moved on to El Paso at the request of the local council and again claimed far-off rain as his own, even though the wind had been blowing in a different direction that time, and in his final session elsewhere in Texas declared success once more, even though it had actually been pouring with rain while his crew set up and launched the equipment.

By this time, judging from contemporary newspaper reports, people were losing faith in Robert St George Dyrenforth. He'd cost the government and local authorities nearly twenty thousand dollars in equipment alone, a startling sum in those days, and especially in a time of such hardship. Newspapers began to refer to him as 'Dryhenceforth' and poems appeared mocking him and his methods.

This is where the slighted Dyrenforth disappears into the quieter life of a lawyer, but thanks to him the cult of rainmaker was underway. Was Dyrenforth nuts? No. Was he a charlatan? His behaviour would certainly suggest so, although there's enough in his story to suggest that, at least to begin with, he had some faith in the potential success of his theories.

He wasn't the only rainmaker in operation, however. While Dyrenforth had attained some sense of legitimacy through government sponsorship, Frank Melbourne was more of a freelance operator. Born in Ireland, Melbourne had emigrated to Australia where he claimed to have perfected his pluviculture techniques in New South Wales. He arrived in the US in the late 1880s offering his services to anyone who could come up with his $500 fee.

Melbourne differed from Dyrenforth in that he attributed his success to a secret mixture of chemicals that he released into the atmosphere through a specially constructed portable machine he called The Baby, which travelled everywhere with him. Small enough to carry in a knapsack, Melbourne's machine worked by extending a twelve-foot-high pipe and cranking a handle to mix and release the fumes rising from his potion. Like most of the rainmakers Melbourne was a show-man and his brief career in the public eye saw him advertised as the star attraction at county fairs across the United States. He was tall, handsome and a dandy: his shirts were hand-made and monogrammed. He first came to public attention in Canton, Ohio, in 1891, when he took bets on whether he could make it rain on a particular day and cleaned up to the tune of several thousand dollars. Soon he found fame around the Midwest, notably in Kansas, Oklahoma and Nebraska, where his routine became well-rehearsed. He'd arrive in town by train in a whirl of publicity greeted by large crowds. He would have sent ahead instructions for the building of a special shed of two storeys with a hole in the roof for the release of his mixture, sur-rounded by a twenty-yard exclusion zone that was patrolled by his brother to keep prying eyes away. While Melbourne claimed some successes – in Cheyenne, Kansas, his appearance coincided with a terrific thunderstorm the likes of which hadn't been seen in years and in which several head of cattle were killed by lightning – he scored more misses than hits, but before high-tailing it out of town with a flurry of excuses about the conditions not being right he'd sell chemicals and copies of his equipment to the locals. As well as the hawking of magic chemicals, most of his money was made courtesy of his brother

Will, who acted as bookmaker, taking bets on the outcome of Melbourne's work.

While he did apparently succeed occasionally in wringing rain from a cloudless sky – one Kansas newspaper reporting on his visit declared, 'Lo! The heavens were suddenly overcast and in the beautiful language of the woolly west, there was a devil of a rain and the Lord wasn't in it' – it didn't take long for the press and public to turn against him, and by mid-1892 reports of his appearances had ceased. In 1894 his body, identified by the monogrammed FM on the breast of his dressing gown, was found in a Denver hotel room.

Melbourne had all the hallmarks of the classic snake-oil sales-man: he knew people wanted it to rain, knew that they'd pay for it to rain and, in the days before mass communication, knew that if he kept moving quickly enough his reputation might stay intact. Claiming early successes and that the development of his methods had taken place on the other side of the world was a good move – it was a long way from Australia to the Midwest – and, like Dyrenforth, if Melbourne was a master of anything it was PR and self-promotion. He'd even claimed to have been run out of Australia after creating so much rain as to cause cat-astrophic flooding. His doubters argued that all Melbourne was doing was loading the law of averages in his favour: some main-tained the days on which he operated coincided with days when rainfall had been predicted in a popular farmers' almanac of the time, but the story of Patrick Murphy shows that such publications were not exactly reliable to start with. However he operated, we know little of Melbourne and will probably never know more. He is as elusive as the strange gas he pumped into the air. The only definite accounts we have come from reports in newspapers of his appearances, and it's unlikely that Frank Melbourne was even his real name. We'll most probably never know where he came from, who he really was or what drove him to take his own life in an anonymous Colorado hotel.

The most enigmatic of the rainmakers was undoubtedly Charles Malory Hatfield. With the likes of Dyrenforth and Melbourne it's fairly easy to speculate as to the sincerity and effectiveness of their motives and operations. With Hatfield, however, the lines are a little more blurred. Of all the pluvi-culture merchants Hatfield is the hardest to shoot down in righteous, mocking flames.

He was born in Fort Scott, Kansas, in the summer of 1875. According to legend his birth coincided with a massive storm that turned the streets into a sea of mud and caused the railroad to come to a complete halt. His childhood was peripatetic as his father moved the family from boomtown to boomtown, making money on the rapid rise in value of properties until eventually settling in California on an estate with an apricot grove near Hollywood.

A snappy dresser with a quietly calm personality, Charles was a natural salesman and became a successful agent for a Los Angeles sewing-machine company. He also possessed a keen intellect and would spend most of his free time in the public library. California was in a depression in the 1890s due to a series of droughts, and times were so hard that the suicide rate among farmers was rocketing. At this time, and in the course of his reading, Hatfield came across Robert St George Dyrenforth and became an instant fan, turning his mind to possible ways in which he might create his own rain. One day while watching a kettle boil he noticed that the steam emerging from the spout

seemed to attract water vapour from a pan bubbling on the
nearby stove. From there he wondered whether if one heated
the right chemicals so that they rose into the sky they could
attract rainclouds in the same way.

In 1902 California entered its sixth year of drought, and
Hatfield had by process of elimination identified twenty-three
chemicals that he believed would create the right conditions for
rain when mixed, heated and sent up into the atmosphere. At
the beginning of April he put his theories to the test on the
family estate. It was a cold morning when he rose early and
began hauling bottles of chemicals and trays up a ladder to the
platform of the ranch's windmill where, once settled in his lofty
position, he mixed his potion, turned on an electric heater
placed under the trays and waited to see what would happen.

Almost immediately he noticed a strange fog drifting towards
the tower. By lunchtime it was raining and it continued for
more than an hour: three one-hundredths of an inch, according to Hatfield's gauge. It wasn't much, but it was definitely a
start. The following month the same experiment induced four
one-hundredths of an inch, and with some more tinkering with
the chemical ratio a month after that the figure rose to sixty-five
one-hundredths.

By the end of January 1903 the drought was so bad that the
clergy announced a day of prayer in an attempt to bring rain. It
didn't work.

At the same time, further experiments had persuaded Hatfield that there really might just be something in his theories.
He told his boss at the sewing-machine company about his
results, who in turn persuaded local businesses to put up a fifty-
dollar prize if Hatfield could use his chemical method to

produce an inch of rain over a five-day period. According to
Hatfield's brother Paul, who was his assistant and confidant, he
did it in two days, two hours and ten minutes. Rain sloshed
down from the sky across a wide area: in some places the water
ended up knee-deep. Some attributed the rain to the power of
prayer, others to Hatfield, as the age-old conflict between science and religion found a new outlet. 'Whether in answer to
the prayers of the church, as a result of Rainmaker Hatfield's
machinations or from natural causes, rain began falling,' pon-
tificated the *Los Angeles Herald*, which would go on to nickname
Hatfield the 'Cloud Compeller'.

He was pooh-poohed as much as he was praised, but took
things in his stride. 'I do not fight nature,' he said enigmatically,
'I woo her by natural means.' On another occasion he stated, 'I
do not make rain. That would be an absurd claim. I merely
attract the clouds and they do the rest.'

He produced even better results the following year, and his
fame began to spread beyond Los Angeles. When he successfully
predicted that eighteen inches of rain would fall between
December 1904 and March 1905 as a result of his chemical
engineering, the *San Francisco Examiner* announced that '[t]here
is nothing about Hatfield to suggest the eccentric, fantasist or
egotist'. He just went about his business quietly and confidently,
responding to critics by saying, 'Censure and ridicule are the
first tributes paid to scientific enlightenment by prejudiced
ignorance.'

So well known did Hatfield the rainmaker become in
California that people adopted the word 'hatfielding' in place
of 'raining', and the Hatfield umbrella became the fashion
accessory no citizen could be seen without. Becoming more

ambitious, Hatfield told *Titbits* magazine that he was talking to the British government about a possible deal to rid London of its 'pea-soup' fogs and there was even mention of irrigating the Sahara. An invitation from Canada produced questions in the Canadian parliament as to whether Hatfield's tinkering with nature might lead to 'the complete submersion of this continent'. An arrangement with the South African authorities to help relieve a grievous drought in the Western Cape collapsed when a letter from the US Weather Bureau informed them that, as far as they were concerned, Hatfield was 'practising deception'.

Either way, he continued to record success after apparent success from Oregon to Texas, but it was a call in 1915 from San Diego that truly sealed Hatfield's fame. In early December 1915 San Diego's water supply was dangerously low. After a meeting with Hatfield the city council offered him a whopping ten thousand dollars if he could fill the massive Morena Reservoir by the end of the following month. Hatfield and his brother set up camp on the northern side of the reservoir on New Year's Eve and raised the twelve-foot tower on which the still secret mixture of chemicals would be heated. Announcing to Paul that 'I'm going to give them the works', he doubled the usual amounts in the trays. Within four days the temperature had cooled noticeably, clouds began to gather and there was rain. Indeed, by 10 January there was a deluge that lasted for days until the run-off into the reservoir was up to 1.4 million gallons per hour, a figure that had more than doubled within two weeks of Hatfield starting the process. In San Diego itself the streets were filled with mud as the rain continued to fall, and by the 16th people were calling for Hatfield to stop. Baseball fields were flooded and the much-heralded San Diego Expo was on the point of being cancelled because of the rain. Telegraph and telephone lines went down, at least one hospital had to be evacuated and the Santa Fe railroad became impassable. San Diego was effectively cut off from the world. Houses started to float away and cattle drowned. By 24 January the reservoir held ten billion gallons of water.

After a brief hiatus, Hatfield renewed the chemicals and a fearsome storm blew up almost immediately. Water gushed down from the hills into the city and San Diego began to make plans to evacuate the areas most at risk. By the morning of 28 January there were more than fifteen billion gallons of water in the Morena Reservoir: it had nearly reached full capacity, and the water level was just two feet from the brim. Rain gauges showed that seven inches of water had fallen in the previous twenty-four hours – sometimes the entire rainy season produced only five inches. However, the full reservoir had come at a price: all but two of the 112 bridges between the reservoir and the city had been destroyed; dozens of people had been swept away and drowned; and the bloated bodies of countless animals floated on the surface of the water.

Focusing on the job at hand, Hatfield had no inkling of what was happening in the valley below the reservoir. The first hint they had up in the mountains that all wasn't as it should be was a message that reached the reservoir supervisor: 'We're on our way to lynch him. His job is done and so is he.' It was then they learned of the effects of the rainmaking. It took the brothers four days to make their way back to the city, where they were astounded at the devastation.

Despite the apparent success of Hatfield's activities, the city council refused to hand over the ten thousand dollars. Crucially, there had been no written agreement and the authorities feared that if they sanctioned the payment they were admitting that the deluge had been deliberately caused and they would be vulnerable to massive claims for damages. Without a legal leg to stand on, Hatfield left San Diego empty-handed and burning with resentment.

He recorded further successes across the western United States in the following years, including causing another flood – this time while experimenting at the edge of the Mojave Desert – but with the Boulder (later Hoover) Dam irrigating southern California since it opened in 1936, droughts were proving less of an issue. During the Dust Bowl of the 1930s there were calls for Hatfield to return to pluviculture but, badly affected by the San Diego incident, he had eventually gone back to selling sewing machines, choosing to live out his remaining years quietly with his scrapbooks until his death in January 1958. He never forgave the city of San Diego for not honouring their deal, saying as late as 1957 that 'the rain of 1916 was an act of Hatfield, not an act of God'.

Could Hatfield have been the real deal? Certainly his successes seemed to far outnumber his failures, and such a biblical deluge as the San Diego flood of 1916 would have to have been a hell of a coincidence otherwise, but as he took his chemical secrets to his grave we'll never know for sure. Charles Malory Hatfield wasn't crazy and I'm pretty sure he wasn't a charlatan. He may have been a successful purveyor of sewing machines but he was not a natural showman, preferring to leave most public matters to his agent Fred Binney. While he did what he did for money, there was a sense of morality and responsibility about

Hatfield. Even his methods were unspectacular: while Dyren-forth's pyrotechnics could hardly be missed and Melbourne adopted every showman's trick in the book, Hatfield preferred to work quietly, well away from the limelight. His confidence in his ability and techniques was unshakable, but while his secrets may have been known to his brother nobody could persuade him to reveal them, even in retirement. 'It is a work that should never be in private hands,' he told one interviewer. 'It is a work that should be controlled by government regulation for the results obtained are so great and the influences spread over such a vast territory that the handling of these important operations, especially in years of drought, is far too great a responsibility for any individual shoulders to bear.'

Maybe he didn't find a government he could trust with his secret formula; maybe it was simply that nobody asked. But Charles Malory Hatfield remains a weather enigma, and the closest any individual has come to mastering the elements.

If Hatfield was a success – if ultimately an unhappy one – perhaps the saddest tale of these meteorological flibbertigibbets is that of Wilhelm Reich. Like the others, Reich wasn't a full-time or trained meteorologist although he was arguably closer to science than any of them. He was an Austrian-born psychoanalyst who in his later years came up with a theory that sounds completely crackers (yet still has supporters today), but according to those who witnessed an incident involving Reich, blueberries and a machine he called the cloudbuster, he produced rain where previously there had been none.

He was born in 1897 and the incident that appears to have defined, or at least greatly influenced, his life occurred when he was twelve years old. Young Wilhelm was schooled at home, and before long it became clear to him that the tutor and his mother were having an affair. When the affair was eventually exposed his mother committed suicide, clumsily, slowly and painfully by drinking bleach. By 1915 his father had also died and when the Russians invaded Wilhelm joined the Austrian army and served until the end of the First World War. Once the war had ended he enrolled at the University of Vienna, met Sigmund Freud the following year and, under his mentorship, became a psychoanalyst.

He moved to Berlin in 1930 but fled to Scandinavia after Hitler came to power and an article had appeared in the press denouncing him as a communist and a Jew. He settled in Oslo in 1934 and it was in Norway that he began to develop his theory that the orgasm is at the root of human existence: that it acts as the regulator of our emotional energy, and hence the better our orgasms the healthier we become. As an energetic shagger himself – he had a bit of a problem with monogamy – Reich was definitely putting in the research. In connection with this, he began to break the taboos of psychoanalysis: he would answer patients' questions directly, sit next to them rather than behind them and, the ultimate transgression, physically touch them in an effort to relax them. As a result, the press turned against him and so he moved on again, setting up in New York in the autumn of 1939.

It was in New York that Reich made the discovery that would define the rest of his career. He went deeper into his theorising into the power and importance of a jolly good orgasm and realised that it all derived from what he termed 'orgone'. Orgone was the primitive cosmic energy that defines us all and also everything from the colour of the sky to the control of the weather. He claimed it was all around us and that he could see it (it was a blue colour, apparently). Orgone was a combination of two elements, which mankind had divided into the ether and God. Put them both together, he said, and you've got orgone, and orgone is very, very good for you. Every illness, from a bit of a sniffle to rampaging cancers, was caused by nothing more than a disruption in one's personal orgone, which could easily be put right.

In 1940 he commenced production of the orgone accumulator, essentially a box made of a combination of materials that concentrated orgone and delivered it to the body, improving health and treating illness. The next step in Reich's mission was to build his cloudbuster. This was a machine consisting of several long metal tubes connected to some hosepipes that led to a container of nothing more complicated than water. It looked like a pipe organ attached to a milking machine. What the cloudbuster did, said Reich, was manipulate streams of orgone energy in the atmosphere to form clouds and induce rain.

This is all very well, you're thinking, but water? And where do the blueberries come in? Water was the purest and best conductor for orgone, according to Reich, and although he claimed several successful tests for the cloudbuster it took a direct request for help to give it its first public outing. In 1953 Maine, where Reich had finally settled at a ranch he named Orgonon, was suffering a severe drought and the local blueberry farmers were seriously concerned about losing their crop. Having heard about Reich's machine they approached him and asked if he would create rain for them in return for a decent-sized fee if he was successful. Reich jumped at the chance and on the morning of

8 July set up his cloudbuster near the Bangor hydroelectric dam. He ran it for just over an hour, and then waited. Nothing seemed to happen at first, but that night it rained for several hours and the blueberries were saved. It could have been coincidence of course, but one eyewitness described how directly above the cloudbuster formed 'the queerest looking clouds you ever saw'.

From this high, however, Reich's career took a plunging swallow dive. The US authorities were concerned that with his talk of orgasms, orgone and propagation of the orgone accelerators Reich was at best committing major fraud and at worst running some kind of sex cult. A Food and Drug Administration investigation obtained an injunction preventing Reich moving his accumulators and his many written works across state lines but in May 1956, while he was away experimenting with the cloudbuster, his assistant moved some accumulators and literature from Orgonon to New York and both men were arrested. The remaining accelerators were smashed, his publications were burnt and Reich was sent to prison for two years on a charge of contempt of court in breaking the injunction, where he died in 1957.

Orgonon is now a museum dedicated to Reich's life and several cloudbusters are on display. One even looks over his tomb in the garden. His orgone theories still have thousands of followers and on YouTube there are even instruction videos on how to make your own cloudbuster (although none are as impressive as the distinctly steampunk one in the video for Kate Bush's 'Cloudbusting', which is based on Reich's story).

The history of weather is littered with characters like Murphy, Dyrenforth, Melbourne, Hatfield and Reich. It could be they all had good intentions; it could be that they were actually right. What all of them had in common was a refusal to accept that man couldn't reduce the weather to logic and master it to the point of control. Everything else, it seemed, was being explained and made to obey, from human behaviour to electricity, so why not something that was as important yet everyday as the weather?

Being men, of course, the routes they took were sex, mixing up potions and making very loud explosions. Crackpots, charlatans and dreamers they may have been, but when it came down to it maybe it was just a guy thing.

15

CAUTION TO THE WINDS: HOW MOVING AIR CAN BLOW OUR MINDS

For all the efforts of scientists, philosophers, snake-oil merchants and the completely cuckoo, while we've been able to categorise, define and even predict the weather we have never managed successfully to impose ourselves upon it. Climate is a little different: we've imposed ourselves on that to the extent that our industrial activity has hastened a process by which it seems there will be palm trees in Regent's Park, cacti in the Great Glen and the good folk of the Faeroe Islands will be producing a cheeky Cabernet Sauvignon before too long. But when it comes to everyday weather we'll never be able to stop the wind with a wave of the hand, or click our fingers and make it rain. Huff, puff and posture as we might, we'll never bring the weather to heel.

The weather, meanwhile, has *carte blanche* to do whatever it likes with us. It can burn us, drench us, blow maps and newspapers into our faces for the amusement of onlookers and make us slip, slide, skid and land flat on our backs like Charlie Brown whenever Lucy whips the football away. The weather can humiliate us and it can also toy with us in other ways: sometimes it just seems to want to do our heads in by making it rain weird things. Frogs and fish are the most commonly observed of this phenomenon, but there have also been instances of it raining blood, nuts, wheat, snakes and ducks. Winkles and a solitary hermit crab fell on Worcester in 1881, while as recently as 2000 there was a brief shower of sprats in Great Yarmouth. Most spectacular of all, in 1987 a shower of pink frogs landed on Stroud in Gloucestershire, a rare albino species native only to Spain and North Africa. There are plausible and rational explanations for these occurrences – tornadic waterspouts lifting creatures into the sky, carrying them across large distances and depositing them on an incredulous populace, for example – but I prefer to think that it's nature just mucking about and reminding us who's boss.

We've already seen how early thinkers used the weather to try to prove they and their people alone were a haven of advanced, intelligent civilisation surrounded by hordes of uncouth, randy, booze-addled, bollock-scratching layabouts, and while such theories are clearly nonsense the weather can and does have a tangible effect on our internal selves as well as our external. Seasonal affective disorder is a well-known and widespread condition that spikes suicide rates during the darker months, but when it comes to specific weather phenomena affecting our minds as well as our bodies then we need look no further than the wind.

As Robert FitzRoy himself put it in his *Weather Book*, 'Air presses on everything within about forty miles of the world's surface, like a much lighter ocean, at the bottom of which we live – not feeling its weight because our bodies are full of air, but feeling its currents: the winds.'

We are closely related to the wind. As FitzRoy says, we have so much air in our bodies that we don't feel the weight of it outside ourselves until it moves. The wind is an invisible relative, its component air keeping us alive, its movement a major contributor to the development of humans and nations. It ruffles our hair, roars in our ears, forces tears from our eyes and whips words from our mouths to carry them far away; it's taken us all around the world and given us our daily bread. The wind can also penetrate our minds, and rarely in a good way.

There is a 1925 film starring Lillian Gish, called *The Wind*, which is based on a late-nineteenth-century novel of the same name by Dorothy Scarborough. It's hard to find these days as, being one of the last silent films ever made (in fact, it was *the* last silent produced by MGM), it was quickly engulfed by the era of the talkie and hence never attained the popularity it unquestionably deserved. It's a brilliant, exhausting, tense, emotionally taut film in which Gish – who was also the creative force behind the project – plays Letty, an innocent girl from Virginia who travels to west Texas to live with her cousin at Sweet Water. The name is entirely misleading as Sweet Water is a godforsaken shamble of a few shacks constantly assailed by a hot, dry, sandblasting wind. On the train at the beginning of the film a fellow passenger, who returns at the denouement to fulfil the role of the bad guy, tells her that the perpetual wind in that part of the world 'drives people crazy – especially women'. Letty looks out through the window at the featureless dust storm and the fear appears instantly in her eyes; that same wind defines a gut-wrenching tale of insanity, rape and murder. MGM insisted on giving the film a happy ending – the novel and original screenplay had Letty disappearing into the wind to her certain death – but even then, when she looks to be living happily ever after with the hero of the film, the final shots are of Letty, who all along has been hypnotised, obsessed with, terrified of and driven mad by the wind, throwing open the door and standing ecstatic with her eyes closed and arms outstretched, embracing and being embraced by the wind.

The Wind was made in extreme conditions in the Mojave Desert, where the temperature was never less than 120° Fahrenheit. Eight aircraft engines created the wind, adding a flushed, weather-beaten authenticity to Gish's extraordinary performance in which she deteriorates from a buttoned-up, smiling innocent on a train to a wild-haired lunatic with wide, haunted eyes that grow wider and darker the more insane she goes in the face of the wind. In a way she *becomes* the face of the wind.

A constant strong wind can be an eerie experience. Most of us have lived or at least stayed in a building around which the wind moans, its mournful pitch rising and falling with the strength of the gusts, but generally for us it's a temporary thing that has gone by morning. Some winds, however, stay for days or even weeks, having a direct effect on the people in their way. *The Wind* may have been an extreme example of how the human psyche can be altered by wind, but it seems there's more than a little truth in the film's central premise.

Voltaire, never in the greatest of health himself, was convinced of the potentially damaging propensities of the wind. 'This east wind is responsible for many cases of suicide,' he wrote of the wind that blows into London from the Thames Estuary. 'A famous court physician to whom I confessed my surprise told me that I was wrong to be astonished, that I should see many other things in November and March, that then dozens of people hanged themselves, that nearly everybody was ill at those two seasons and that black melancholy spread over the nation, for it was then that the east wind blew most consistently.

'This wind is the ruin of our island,' he went on. 'Even the animals suffer from it and have a dejected air. Men who are strong enough to preserve their health in this accursed wind at least lose their good humour. Everyone wears a grim expression and is inclined to make desperate decisions. It was literally in an east wind that Charles I was beheaded and James II deposed.'

Hippocrates, the father of medicine, was also convinced of the wind's effect on health. 'In towns frequently exposed to winds,' he wrote in *On Airs, Waters and Places*, 'such as those which blow between east and west and which are sheltered from the north winds, the slightest cause can turn sores into ulcers. The inhabitants lack force and vigour, the women are sickly and barren, the children are attacked with convulsions or sacral disease and the men are subject to dysentery and long fevers in the winter.'

While the effects of the wind may not be quite as bad as these two make out, there is scientific evidence to suggest that as many as 30 per cent of us are sensitive to the winds, especially those particularly cold or warm ones specific to certain places; winds like the Mistral, the Sirocco and perhaps the most famous of the sickly winds, the Föhn. These winds are born of the channelling of air streams by topography, notably hills and mountains, making them unique to a particular locality. As the air moves at speed across the land it can find itself pushed upwards by a range of mountains. Finding itself suddenly compressed between the peaks and the ceiling of the atmosphere, the air is squeezed over the top and spat out the other side in the form of a persistent, violent wind that picks up pace as it descends the slopes. In some cases the air also warms up as it sinks, and it's this process that causes the Föhn to blow southerly across the northern valleys of the Alps into Switzerland and Austria with often devastating effects.

The Föhn is an eerie wind: it's known to make people's hair become alive with static and cracks can appear in the walls of houses and in furniture. The Föhn can propagate fires: the Swiss town of Glarus was almost completely destroyed by a Föhn-related conflagration in 1861, while Meiringen – the town at the Reichenbach Falls, scene of the demise of Sherlock Holmes – all but burnt down twice in similar circumstances in 1879 and 1891. In January 1919 whole swathes of mountain forests around St Gall and Appenzell were destroyed by fires sparked by the hot wind; in some places it is sometimes necessary to ban smoking and cooking as the Föhn passes, with many communities appointing special *Föhnwächter* – 'Föhn watchers' – to enforce these temporary but essential rules. So warm is the Föhn that in twenty-four hours it can melt snow that would otherwise take a fortnight, while on a lake in Appenzell in 1863 ice twelve inches thick melted in one day. In a way it's as if the Föhn carries instant spring in its pocket, although crops and fruit can be damaged: stories abound of apples being 'cooked' while still on the trees.

People are affected by the Föhn in more ways than just laying off the Silk Cut and making do with sandwiches. Medical research has shown that something about the Föhn pushes up serotonin to abnormally high levels, increasing the chances of stress, insomnia, aggression and migraines. People feel more irritable than usual and the number of traffic and industrial accidents rises by as much as 10 per cent. Heart attacks are more common, as are rheumatic pains, epileptic seizures and asthma attacks; some hospitals in Switzerland even postpone surgery during the Föhn as the risk of bleeding out and of thrombosis is higher than usual. Many scientists believe that the wind creates positive ions in the atmosphere, which in turn has a negative effect on those people sensitive to them.

The Mistral is formed in a similar way to the Föhn, but is bitingly cold. It originates in a cold front that still feels bitter even allowing for the warming of its compression. Most common in the spring and the autumn when the air pressure in northern France is high while over the Mediterranean it's low, this icy wind hurtles along the Rhône valley, which effectively acts as a wind tunnel. Railway carts have been known to topple in the teeth of the Mistral, people are thrown from horses and in the nineteenth century coaches would be secured by ropes while everyone hunkered down until the wind had passed. In the 1920s some badly braked railway trucks were pushed fully twenty-five miles from Arles by the force of the wind. As long ago as the time of Christ, Strabo had the misfortune to experience the Mistral, describing it as 'an impetuous and terrible wind which displaces rocks, hurls men from their chariots, breaks their limbs and strips them of their weapons and clothes'. The Mistral can last up to a week, drying out soil and ruining crops – the Provence landscape is lined with densely packed cypress trees and reed screens placed by farmers in an attempt to prevent the wind stripping the moisture from their land and laying waste to their crops. Towns are laid out in order to reduce the effects as much as possible: roads are planned at right-angles to the direction of the wind to avoid creating further tunnels, while most doors and windows are placed on the sheltered south and east sides of buildings.

Its sister wind is the Bora, a fearsome, cold wind that rushes down from the Alps towards the coasts of Istria and Dalmatia and may well have taken its name from Boreas, the Greek god of the north wind. The Bora is a wind so violent that in the port city of Trieste there are railings on walls and heavy chains on thoroughfares for pedestrians caught in the wind to cling on to, so that they are not blown away. When Stendhal visited Trieste in 1831 he said, 'I call it a high wind when I hold on to my hat and a Bora when I'm in danger of breaking my arm.'

Europe has many of these localised winds, all of which have curious names handed down from history: the Haizbeltza, or 'black wind' of the Basque Country; the Leste, a hot easterly in the Canary Islands; and the scorching, dry Xlokk in Malta, to name but three. Even Britain has its very own named wind, although you may not have heard of it. Blowing north-easterly across the southern escarpment of Cross Fell in Cumbria (where the prevailing winds are from the south-west), the Helm wind can trumpet away for anything from a few hours to up to three weeks. Possibly taking its name from the 'helmet' of cloud that forms over Cross Fell to signal the coming of the Helm, it only passes over a relatively small area and runs out of puff before it's gone more than a handful of miles, but the Helm can cause devastation similar to the Bora. There are no buildings on the fellside for a reason, while the houses further afield are built facing away from the origin of the Helm. It is strong enough to blow sheep off their feet as well as warm and persistent enough to trigger the detrimental health effects of a regular Föhn on any people in its way, but the Helm is sufficiently small that it remains localised, keeping its consequences to a minimum. Its parochialism and strange civility – it obligingly blows itself out just before reaching Penrith, the nearest major town – make it an oddly appealing, some might say very British, weather phenomenon.

Europe, with its undulating topography and relatively warm seas, is a nest of such winds, winds that cut about the continent invisible and unobserved while going about their malevolent business. Winds that fly in all different directions, some hot, some bitterly cold, swooping down mountainsides and racing through valleys bringing migraines, palpitations and snow or dust, stealing moisture and immolating crops in a continent-wide, chaotic yet perfectly organised dance of constantly moving air.

While we have never controlled the winds, it has not been through want of believing it can be done. As recently as the early twentieth century there was a woman in Stonehaven, just south of Aberdeen, whom sailors would consult before heading out to sea in the hope she would bestow good weather fortune on them. In return for bags of coal for her fire she would hand over wind charms, pieces of red thread tied with three knots: one to be untied for a light wind, two for a strong wind and the third ... well, the third was never to be untied for any reason. If any fisherman did untie all three they never came back to talk about the consequences. The French explorer Pierre Martin de la Martinière had a similar tale from his expedition to Lapland in 1653: his ship was becalmed for several days when the winds dropped. A local Sami wise man sold the explorer wind knots in return for money and tobacco and, sure enough, on undoing the first knot a breeze picked up to set them on their way. Later in the voyage the wind dropped again and de la Martinière untied the second knot with the same result. When it happened a third time a howling gale blew up, a horrendous northwesterly that looked set to dash the ship against the coast and was only defeated by an intense session of deck-top prayer by captain and crew.

More than a century earlier, around 1800, Sir Walter Scott had visited a wind witch at Kirkwall in the Orkney Islands. 'We clumb by steep and dirty lanes, an eminence rising above the town,' he recalled. 'An old hag lives in a wretched cabin upon this height. She subsists by selling winds. Each captain of a merchantman, between jest and earnest, gives to the old woman sixpence. She in return boils her kettle to procure a favourable gale.'

The weird sisters encountered by Macbeth gave each other winds as tokens of friendship, while Gustavus Adolphus of Sweden, father of Descartes's nemesis Queen Christina, credited the wind magic of the Sami in his armies for their contribution to his success in the Thirty Years War. They, as Pierre Martin de la Martinière could confirm, would seem to have been specialists in that kind of thing.

It's not just on the sea that we've managed to make use of the wind: the windmill, for example, is one of our greatest inventions. The cover of the Ladybird book I'd read in South Wales had one on its cover as an appropriate illustration of how we have harnessed the weather for our needs.

The catalyst for the introduction of windmills to Britain was the Crusades. Early windmills had appeared in sixth-century Persia in the form of square towers with horizontal sails that spun like a child's roundabout. They are called panemone windmills, and were originally used for pumping water and eventually to help grind corn. By the twelfth century the idea had spread as far as China. It is thought returning Crusaders brought the idea back with them from the Middle East. There are no records of any windmills anywhere in Europe at the end of the eleventh century, but within a hundred years so many had sprung up that a new papal tax was introduced, specifically applying to them. The first recorded mention of a windmill is in Normandy in 1180, when an abbey at Montmartin-en-Graignes received the gift of a piece of land located 'close to a windmill'. Eleven years later the windmill had reached Britain, the first mention being of one constructed at Bury St Edmunds in 1191. It didn't last long, as the local abbot, who had the monopoly on grinding corn, soon insisted it be demolished.

The oldest surviving windmill in Britain, at Bourne in Cambridgeshire, dates back only as far as the seventeenth century, but there are a number of medieval carvings and illustrations to show that windmills were widespread, both simple pole structures and the tower type that a generation of us grew up watching Windy Miller walk in and out of, apparently oblivious to the turning sails that came within inches of swatting his head clean off his shoulders in *Camberwick Green*.

Today we are much more likely to see wind turbines, the graceful white sentinels that puncture horizons, both lonely singletons and in vast windfields, turning silently as we pass. They look like the very epitome of modernity, sleek and elegant, utilising a basic technology that dates back hundreds of years in order to turn wind into energy. The first instance of the wind being harnessed to produce electricity dates back to 1887, when a Scottish engineer and university lecturer named James Blyth powered the lights of his holiday home in Marykirk, Aberdeenshire, via a dynamo connected to a 33-foot-high tur- · bine in the garden. Flushed with success and public-spirited zeal, Blyth approached the burghers of Marykirk offering his technology to light the streets of the village but they turned him down on the entirely reasonable grounds that electricity was the work of the devil. He did manage to provide power to the lunatic asylum at Montrose via a windmill in the grounds, but Blyth's invention wasn't seen as economically viable in the UK. The baton was eventually picked up by the Danes and the Americans and, with concern about global warming and the finite supplies of fossil fuels, the twenty-first century could well see the wind become the major source of energy, returning the yoking of the winds to the heart of our existence.

What would Francis Beaufort make of it all? The man who spent his career hitching himself to the winds on the ships that took him around the world and, ultimately, defining them?

The journal Beaufort began that night in January 1806, when he first set down his wind scale, would go on to be filled with weather records, as would many more that came after it. The records he started in that contented postprandial glow began a regime of weather recording that would continue almost daily for the rest of his life. When he became hydrographer to the Navy in 1829 it solved his financial worries and was at last a post commensurate with his stature and reputation. It had taken until the age of fifty-five, but finally he'd overcome nepotism and privilege to take his rightful place at the pinnacle of his profession. It was also a job that meant the Beaufort scale could be disseminated throughout the Navy and, by extension, to the rest of the maritime world. Not only that, Beaufort lobbied for meteorology to become a much larger concern to the Navy, and eventually barometers were issued to all captains at his behest. Despite this late success, Beaufort still found it hard to find true happiness. Heavily involved with the Arctic Council, he'd had reservations about his old friend Sir John Franklin undertaking his voyage in search of the Northwest Passage in 1845. When Franklin disappeared Beaufort blamed himself and took personal charge of assigning missions to attempt to find and rescue him. Franklin was never found.

In 1855 Beaufort stepped down. Although exhausted from a lifetime of effort both at sea and behind a desk at the Admiralty, he took to retirement badly, finding it hard to relax. He kept up his weather journal (spicing it up with a concurrent record of the regularity and consistency of his stools) but his days passed slowly and his health deteriorated, even after moving to the more pleasant surroundings of Brighton where he could at least watch the sea. He died at home there on 16 December 1857 and is buried in Hackney in a simple white tomb along with his wife and daughter, sheltered from the wind that blows cold through the East End. His genius wind scale is still in use today, even in this age of technology: a couple of minor alterations aside, it has never been improved upon and never will. For that alone Admiral Francis Beaufort deserves the renown his eponymous scale has brought him. In addition, his name graces the Beaufort Sea in the Arctic and Beaufort Island in the Antarctic, ensuring that his memory extends to the very ends of the earth.

Beaufort once called on William Wordsworth at his home in the Lake District, but the poet had been out at the time. One can only wonder at what a conversation between the father of the wind and the man who found inspiration in the weather to produce some of the finest poetry ever written might have covered: for once, a chat about the weather between strangers might not have been 'the last resort of the unimaginative'. Even so, Beaufort left us with a poetry of his own.

16

FROST FAIRS AND THE SNOWFLAKE MAN:
A STORY OF SNOW

It's probably no surprise that the phrase 'as beautiful as a football stadium' is not in common use. Architects may sometimes purr over a finely turned cantilever roof, but in general the football stadium is not renowned for its pulchritude. Stadiums in the 1980s were even less likely to be discussed in positive terms, and in the league table of descriptive nouns Charlton Athletic's ground would have struggled to heave itself above 'eyesore'. A vast concrete bowl of stepped terracing laid over the remains of an old chalk pit, which undulated as if determined that no two people should be standing at exactly the same height above sea level, The Valley had changed little between the 1930s and the time Wolverhampton Wanderers were due to visit one Saturday afternoon early in 1983.

I'd just about reached an age when my parents would allow me to go to Charlton matches unaccompanied, a privilege I was determined to enjoy as fully as possible. At lunchtime, *Football Focus* presenter Bob Wilson ran through the matches postponed due to the blanket of snow that had been laid over the country during the night and Charlton Athletic v Wolverhampton Wanderers wasn't among them, so I trudged happily off to the bus stop. I was conscious on my way down the hill to the stadium that there weren't many people joining me, and when my usual programme seller wasn't in his customary position I began to fear the worst. Then a man walking towards me with his head down against the cold looked up from beneath an enormous bobble hat whose redness almost matched that of his cheeks and said, 'It's off, mate.' My journey had been a wasted one after all, but rather than just turn round and catch the bus home again I carried on to the ground just to have a look. Outside the main gates a piece of board with the words 'MATCH OFF' painted on it was propped against a chair, but the gates were open. As my shoes munched through the fresh snow I shivered with the thrill of privilege: I was inside the ground, inside the place where my dreams were played out, and hadn't had to pay to get in.

I'd never noticed just how ugly The Valley was back then. The barbed wire, the steel-mesh fencing, the crumbling terracing, the dingy, bare-bulbed tea huts – there were more welcoming industrial estates. I'd never thought it ugly and I'd never thought it beautiful. All that mattered was that it was the place where my heroes played out their fortnightly, frequently fruitless, combats in the bottom half of the Second Division. The vivid green of the pitch held my attention, and when I wasn't looking at that I had my head buried in the match programme.

But that Saturday, as I looked out across the empty stadium I realised that the snowbound Valley was probably the most beautiful thing I'd ever seen. The snow had settled everywhere, unblemished on the pitch except for the footsteps of the referee where he'd inspected the surface and immediately declared it unplayable. It had settled on the barbed wire, on top of the advertisement boards for the likes of Simba Car Alarms and Metaxa brandy, it had settled on the roof of the grandstand, but most noticeably it had settled on the vast bank of terracing that dominated the eastern side of the ground, an expanse so large that in its day it had accommodated forty thousand surging, stamping, singing people on its own. Normally it was a purely functional grey-brown mess of cracked concrete with a couple of thousand hardy souls huddled at its centre. But today wasn't normal. Today was very different. Today it was a mountain face of pure unsullied white: dignified, elegant and quite beautiful.

Silence hung over the place too. Even with the tiny crowds Charlton were attracting in those days there would usually be an energy of anticipation at this time on a match day, the music crackling out of the loudspeakers, the chants and songs from behind the goals, the hubbub of conversation, the greetings of people who'd known each other for years but only here and only in these fortnightly winter doses, the smell of burnt onions and cigarettes, the cries of the programme sellers, the infinite individual movements of the crowd. But that day, with just a couple of dozen people in the place, standing around with their breath disappearing skyward in clouds, there was just a snowy, silent stillness that emphasised how The Valley looked absolutely, breathtakingly gorgeous.

I've seen countless matches there over the years, many if not most of which I have entirely forgotten. But every detail of that snowbound Saturday is fixed in my memory, from the silent grace of the east terrace to the dainty scrunch of the Wolves players' tasselled grey slip-on shoes as they tiptoed gingerly across the snow from the dressing rooms back to their team coach for the journey north.

The snow had made the ugly and functional into something stately and beautiful: it was the ultimate meteorological makeover and a transformation that could only be achieved by snow. Snow brings forth splendour from the sky, puts its finger to its lips and reminds us, through silent contemplation, of the beauty around us.

There are few greater weather thrills than seeing a flurry of snowflakes in the air. Nothing brings offices and homes to a standstill quicker than the cry, 'It's snowing!' It's an echo from earliest childhood as we rush to the window to watch the creation of a natural playground whose potential for magical possibility never leaves us. Snow moves like no other meteorological phenomenon: raindrops and hailstones want to race each other to the ground, fog just loiters there all dopey, but snow descends in its own time. A snowflake billows and corkscrews in a feather-light dance with gravity that ensures it doesn't so much hit the ground as alight on it when ready.

And what a thing is a snowflake! Its formation and construction make up one of nature's most understated processes to bring about its most beautiful creation, but the true, intricate, close-up beauty of the snowflake remained largely hidden until around the turn of the twentieth century and the remarkable work of yet another single-minded, extraordinary man with a rare passion for the things that go on above our heads.

The life of a snowflake begins when water vapour or an extremely cold water droplet in the lower atmosphere forms an

ice crystal around a tiny speck of floating matter: a dust parti-
cle, say, a minuscule grain of salt or volcanic ash – the invisible
detritus that's floating above us at all times. These crystals then
attract more water vapour, vapour that turns straight into ice
without having a liquid phase, producing a snow crystal. The
shape of the ensuing flake depends on the surrounding temperature and the amount of water vapour in the air. At around
freezing, they tend to be needle-shaped. In temperatures as low
as −18°C the classic six-branched snowflake forms; ten degrees
lower and it's a more complete hexagon. As these snow crystals
descend they collide and join with others, and these aggrega-
tions fall to earth as snowflakes.

It is their apparent symmetry that makes them so extraordinary. Chinese scholars had remarked upon it as early as the
second century BC. The thirteenth-century German philosopher and Dominican friar Albertus Magnus was the first in
Europe to write about snow crystals, describing their star shape
but theorising that these particular flakes fell only in February
and March. According to legend, Albertus had always struggled
with his studies (one nineteenth-century account records rather
uncharitably that 'for the first thirty years of his life he appeared
remarkably dull and stupid and it was feared by everyone that no
good would come of him') until one day the Virgin Mary
appeared to him in the cloisters and granted him the gift of
excellence in philosophy. Now bursting at the temples with
philosophical insight and intellectual vigour Albertus, together
with his student Thomas Aquinas, apparently either acquired or
created a potion that contained many of the ingredients of the
elixir of life. They put this to its most obvious use: to bring to
life a bronze statue, which they immediately set to work as a

domestic servant. Unfortunately they had inadvertently given the statue not just the gift of life but the gift of the gab: it wouldn't shut up, until one day it disturbed Thomas when he was at work on some brain-busting mathematical problem. At the end of his tether, Thomas picked up a hammer and smashed the statue to pieces.

Albertus is also the subject of a pretty good snow story. In later life he wanted to build a monastery on a piece of land outside Cologne, which belonged to William, Count of Holland. One winter Albertus invited William and his entourage to dine and William accepted, but when he arrived with his retinue he found that the meal had been set out in the courtyard where the snow drifted several feet deep, piled high by the freezing north wind that blew hard across this curious dining vista. William took one look at the scene and was about to clear off in a huff when Albertus gave a signal, the clouds rolled away, the sun came out, the wind dropped, the snow melted, the trees burst into bud, flowers grew from beneath their feet and the birds started to sing. The whole event passed under a blazing sun, but as soon as the meal was over Albertus gave another signal. The clouds gathered again, a blizzard whipped up, the leaves fell from the trees and the birds dropped dead out of the sky from the cold. Somehow this meteorological sleight of hand impressed William enough to award Albertus his piece of land.

Johannes Kepler was the next man to seriously ponder the derivation of the snowflake, some five hundred years after the alfresco-dining friar. 'There must be some definite cause why, whenever snow begins to fall, its initial formation invariably displays the shape of a six-cornered starlet,' he wrote in 1611. 'For if it happens by chance, why do they not fall just as well with five corners or with seven?'

Kepler went on to point out parallels between snowflakes, honeycomb and the pattern of seeds inside pomegranates, but was unable to posit a theory as to why they formed in such a way. Descartes, who had plenty of opportunity to study snowflakes during his ill-judged and ultimately fatal decampment to Sweden, noticed that branches lead out from each side of the stems of hexagonal snowflakes at an angle of about 60 degrees, meaning the branches themselves are separated by 120 degrees. Descartes didn't make the connection himself, but it's no coincidence that the two hydrogen atoms in a molecule of water branch off the oxygen atom at an angle of around 120 degrees.

The next major step forward in the history of our understanding of snow came neither from a philosopher nor a scientist. It certainly didn't come from a brainy monk with a mouthy metallic manservant either. Instead it came from an otherwise ordinary home-schooled farmer from a valley in the shadow of a Vermont mountain.

Wilson 'Snowflake' Bentley was from the mould that gave us Luke Howard and George James Symons. It was Bentley who gave rise to the notion that no two snowflakes are alike ('Every snowflake has an infinite beauty which is enhanced by knowledge that the investigator will, in all probability, never find another exactly like it,' he wrote in 1922) and his thousands of beautiful black and white microscopic photographs taken over nearly five decades provide a fantastic gallery of snowflake formation for which meteorologists and scientists are grateful even today. He was such a pioneer that even the methods he used to photograph them are still in use.

Part scientist, part photographer, part aesthete, Wilson Bentley was born in Jericho, Vermont, in 1865 to a farmer father and a former schoolteacher mother. Like his brother, he was expected to work on the farm from an early age (something he continued to do for the rest of his life), but it was his extra-curricular activities that would bring him fame and the gratitude of meteorology.

He developed an early obsession with the microscope that his mother had left over from her teaching days and, as Bentley himself said, 'When other boys of my age were playing with pop guns and slingshots I was absorbed in studying things under this microscope.' Most of all, though, he was fascinated by the snow that would arrive in Vermont in November and stay often well into May – 'I can't remember a time I didn't love the snow,' he told an interviewer in 1925 – and he developed a method of catching snowflakes on a black board, quickly transferring them to a microscopic slide and examining them before they melted or evaporated. Captivated by the beauty he was seeing close-up, the teenage Bentley would sketch the hexagonal designs as best he could, frustrated that the full majesty of the snowflake couldn't be reproduced exactly by his pencil and shared before it vanished for ever.

'Under the microscope I found that snowflakes were miracles of beauty; and it seemed a shame that this beauty should not be seen and appreciated by others,' he told the *American Magazine* in that 1925 interview. 'Every crystal was a masterpiece of design and no one design was ever repeated. When a snowflake melted that design was for ever lost. Just that much beauty was gone, without leaving any record behind. I became possessed with a great desire to show people something of this wonderful loveliness, an ambition to become, in some measure, its preserver.'

When he was seventeen, Bentley persuaded his sceptical father to buy him a bellows camera that could be hooked up to a microscope. At a hundred dollars it was a considerable outlay, but Bentley Senior gave in to the stereo entreaties from Wilson and his mother, and the camera, a piece of apparatus that he would use for the rest of his life, duly arrived.

It took a while to master – Bentley had never operated a camera before, let alone one intended for such delicate, intricate subject matter – but on 8 January 1885 he finally captured his first perfect microscopic image of a snowflake. Thus began a lifetime's work that would produce more than five thousand snowflake photographs, all different, all unique, all beautiful.

For a dozen years Bentley toiled away alone in his valley at the foot of a mountain, never thinking for one moment he was doing anything important or groundbreaking. He became a local curiosity – 'I guess people here thought I was crazy, a fool,' he said later – but it wasn't until George Perkins, Professor of Natural History at the University of Vermont, heard about Bentley's work that he first had an inkling his photographs might be of some wider importance. At Perkins's behest Bentley published his first article about the composition of snowflakes in an 1898 edition of *Popular Science Monthly*, the first of many pieces that he would write for publications as prestigious as *National Geographic* and the *New York Times*. In that first piece Bentley posed the wonderful question of snowflakes: 'Was ever life history written in such dainty hieroglyphics?'

Bentley was more than an obsessive photographer: he argued that the different shapes of snowflakes came from different parts of a snowstorm cloud, and that variations in temperatures affected the formation of snowflakes, theories that were largely correct and way ahead of their time. Away from snow, Bentley was the first person to consider the size of raindrops as important in theorising about the formation of rain, catching rain in a dish of flour and measuring the doughballs produced by each individual drop.

Possibly because he was an obscure, self-taught farmer who made his scientific observations between milking cows and fixing fences, few of Bentley's theories were picked up by scientists during his lifetime. It took until 1924, nearly forty years after his first successful photograph, for him to be awarded any kind of research grant, for example, but Bentley wasn't really interested in money anyway. While the farm doubled its head of cattle after he took it over with his brother on the death of their father, Bentley would give lectures for nominal or no fees and sell slides of his pictures for little more than cost price. He was just happy doing what he did, which also included keeping a detailed weather journal, observing the Northern Lights and exploring the mountains as part of an interest in geology. Money didn't matter to him; he lived frugally in a wing of the farmhouse, the rest of the house occupied by his brother's large family. An accomplished musician, he would often sing popular songs of the day at the piano to entertain local kids, but otherwise he kept himself to himself.

In the summer of 1931 Bentley's national fame was at last assured when a book containing more than two thousand of his photographs was published at the behest of William J. Humphreys, Chief Scientist at the US Weather Bureau. Called simply *Snow Crystals*, it featured a praise-filled introductory chapter by

Humphreys and was an immediate success, but Bentley didn't have long to enjoy it. In December of that year, aged sixty-six but with his enthusiasm for snow undimmed, he noted 'cold north-easterly, snow flying' in his weather journal, trudged out into a blizzard with the same camera he'd persuaded his father to buy half a century earlier, was caught in the teeth of the snowstorm and contracted pneumonia, dying at home in bed two days before Christmas.

The obituaries of this simple, brilliant man were published nationwide, but it was his local newspaper that summed him up best: 'Wilson Bentley was a greater man than many a millionaire who lives in the luxury of which the "Snowflake Man" never dreamed.'

It was only later that Bentley's brilliance was truly appreciated. His photographs, each capturing a fleeting individual creation of nature in all its beauty and majesty before it faded to nothing, provide a permanent record of both the science and art of the snowflake. The monochrome slides are hauntingly beautiful, a permanent epitaph for a man who, like the flakes themselves, dazzled briefly and brilliantly before fading away. A remarkable man with a remarkable legacy.

The Gulf Stream means that we are spared the particularly harsh winters experienced by Bentley and our other neighbours on similar latitudes, but that doesn't mean we haven't had our moments, particularly in recent years. The winter of 2009–10 was a particularly cold and snowy one, with temperatures in Manchester dropping to −17°C while at Altnaharra the mercury sank to −22.3°C. Blizzards reduced motorways to a standstill and motorists were faced with the choice between sleeping in their freezing vehicles or abandoning them altogether. The Scottish transport minister confirmed that the roads there had not seen conditions like it in twenty years or more. The Kentish towns of Deal and Sandwich were effectively cut off by snowdrifts. Schools and airports closed across the country and forty-five thousand Scottish homes were left without power. On 20 December two thousand people were stranded in the Channel Tunnel for sixteen hours, when the cold triggered an electrical fault. By Christmas Day twenty-one people were reported to have died as a direct result of the cold snap.

Among them were five people who died in three separate Scottish avalanches, but that figure is still eclipsed by Britain's worst avalanche disaster, which took place in the unlikely setting of Lewes in East Sussex. The winter of 1836 had been harsh across the country, and a combination of blizzards and high winds had left a large accumulation of snow teetering at the top of Cliffe Hill on the eastern side of the town. The snow was directly above Boulder Row, a line of ramshackle cottages that housed some of Lewes's poorest people. Although concerns were expressed at the quantity of snow hanging a hundred feet above the street it wasn't deemed a threat serious enough for people to leave their homes. After all, where would they go? The cottages and their fireplaces were their only defence against the cold.

At around 10.15am on 27 December there was a crack and roar as the snow cornice gave way and thundered down the almost sheer hillside. Such was the force with which the avalanche hit Boulder Row that the houses practically exploded, leaving nothing but a mountain of snow from which protruded pathetic pieces of shattered timber and personal belongings. Seven people were pulled out alive but eight more were killed;

their names are memorialised on a plaque in a local church. Boulder Row was never rebuilt, but a pub stands on the site. It is called, appropriately, the Snowdrop Inn.

The Lewes disaster occurred towards the end of what is known as the Little Ice Age, a period of around three hundred years, up to the middle of the nineteenth century, when the climate in the northern hemisphere was particularly cold. Dickens was a child at its tail end and became used to cold, snowy winters, hence their proliferation in his work and their continued proliferation in our culture today: the snowy scenes on our Christmas cards are a hangover from this period just before the climate warmed again.

Perhaps the best known manifestation of this chillier era, however, is the Thames Frost Fair. Records of the Thames freezing over date back to AD 695, when a market was set up on the ice for the first time. The river would go on to freeze eleven times in the next thousand years (the winter of 1434–5 was an especially harsh one: the Thames froze for nearly three months, and at the height of the frost one could walk along the river from London Bridge as far as Gravesend, twenty-five miles away in Kent; while in 1536 Henry VIII travelled along the frozen river from his palace at Greenwich to central London by sleigh), but it was between the seventeenth and nineteenth centuries that Jack Frost brought the river to a standstill with his greatest zeal. The Thames froze on twenty-three occasions between 1620 and 1814, with 1683–4 a particularly vintage winter for fans of perambulating on frozen water. It was a season that saw temperatures plummet right across Europe: the *London Gazette* said as late as February 1684, 'We have nothing of moment from Flanders, the Weather has been excessive cold there as well as here, to that degree, that not only the greatest Rivers are frozen, but the Sea is covered with Ice for many leagues from the Shore.' At Deal the sea froze to a distance of two miles, while shipping in the English Channel faced serious danger in the form of large blocks of ice at large in the sea. An Oxford academic recorded that winter how the ink in his inkwell froze solid at his fireside 'unheeded of the fire's warmth and the heat of academic passions'.

That year the freezing of the Thames happened on 19 December with alarming suddenness. The river froze quickly enough to catch out people in boats and many found themselves trapped in ice: one man out fishing at Blackwall, unable to reach the safety of the shore, froze to death where he sat.

By the turn of the year the temperature had dropped further and John Evelyn recorded in his diary that on 9 January he 'went across the Thames on the ice, now become so thick as to bear not only streets of booths, in which they roasted meat, and had divers shops of wares quite across as in a town, but coaches, carts and horses passed over'. Known as Freezeland or Blanket Fair, the amount of commercial activity on the ice was extraordinary. Entire streets of booths and stalls criss-crossed the river. There were cock-fights and ox-roasts, bull-baits and puppet shows. A printing press churned out personalised certificates: 'And sure, in former Ages, ne'er was found, A Press to print where men so oft were droun'd' as one contemporary verse described it. Another detailed how 'from a Brandy-Hovel near Black-fryers, the growing Freezeland to a street aspires', a phenomenon so worthy of note that King Charles II himself paid a visit and even bought a personalised souvenir from the enterprising printer.

It wasn't all fun, though: the deep depression over the city laid across it a slab of cold air that prevented much of London's smoke from dispersing effectively. Evelyn wrote of the 'fuliginous steam of the sea-coal' that 'exceedingly obstructed the breast so one could scarcely breathe'. The fishermen of the river lost their income, the docks were useless and hence so were the dockers. The river was the main avenue of the city's trade – it's hard to imagine now just how frantically busy the Thames was right up until the early twentieth century – vital supplies like coal took longer to transport by road, the price therefore went up and the poor froze to death in their hovels.

When the thaw came in early February, it came as quickly as the freezing. A shower of rain escalated the thaw, causing the ice to crack, break up and be carried at speed towards the estuary. Boats were smashed and bridges damaged, but within hours Old Father Thames was back to his old self.

The era of the Frost Fair ended in 1814, the last time any kind of trade or entertainment was provided on the ice. The fair lasted for just four days in February, with an elephant being led across the ice by Blackfriars Bridge, but the warming of the climate and the 1831 demolition of the old, narrow-arched London Bridge that had obstructed the tide and aided the freezing meant the Thames had solidified for the final time. Down sides though the freezing unquestionably had, imagine how much fun it must have been playing football or skittles, or buying a commemorative certificate in the middle of a river.

The recent winters have taken the fun out of snow for many people. 2010–11 in Britain and Ireland was the second severe winter in succession (December 2010 was Britain's coldest on record) and for the first time in my lifetime people generally weren't pleased to see the snow. There were deaths almost Victorian in their tragedy: as well as the increase in car accidents, a man froze to death in a Cleethorpes caravan and another in the grounds of Bangor Cathedral in North Wales. The big freeze was spoken about in economic terms – Britain was losing £1.2 billion a day, according to some reports – and the transport infrastructure ground to a halt again, with people rolling their eyes and citing 'the wrong type of snow'. The snow seemed to just hang around for far too long; the dirty brown mountains at the roadsides refusing to thaw, the pavements and roads remaining as ice rinks long after the novelty wore off, making every journey fraught with risk.

In these austere times when everything, even a cancelled train, can have a financial loss figure attached to it, maybe the magic of snow is fading. After our recent winters the heart is now as likely to sink as sing at the first snowflakes of the winter. That's a terrible shame. For me, though, a glance at some of Wilson Bentley's snowflakes brings that magic right back and I think of his delight at the sight of falling snow; a delight that dimmed not one iota from childhood to old age.

I also think of a small boy standing in a deserted, all but derelict football stadium draped in white and thinking it's by far the most beautiful thing he has ever seen.

17

THE MANY FACETS OF FOG

In 1954 Frankie Howerd was hot stuff. He'd been on the wireless as the star of *Variety Bandbox* ever since the end of the war when he was approached by the director Val Guest to make a film in which he'd have the leading role. Despite Howerd's initial reluctance to grace the big screen, the upshot of Guest's overtures was *The Runaway Bus*, also starring Margaret Rutherford, Petula Clark and the man who later played Charlie Hungerford in *Bergerac*. Howerd played Percy Lamb, a coach driver tasked with taking a small group of passengers from a fog-bound Heathrow Airport to Blackbushe airfield in Hampshire, where the higher ground meant the fog wasn't as thick and so there was more chance of aircraft taking off. A bullion robbery and a deserted village add crime and suspense to the comedy but it's not, in truth, a particularly great film. A low-budget cash-in on the man of the moment, *The Runaway Bus* showcases Howerd's familiar range of world-weary eye-rolling and haughty faux indignation but it's certainly no classic.

The screen presence that makes *The Runaway Bus* relevant to this story is the fog. It's there throughout the film, a constant, silent presence and the *raison d'être* of the entire plot (such as it is). For someone who arrived long after the fifties, *The Runaway Bus* provides for me a fascinating glimpse of just how different the fog was back then. In the early scenes we see Howerd groping his way along walls until he stumbles upon his vehicle, and when he sets off peering into the gloom he becomes immediately, hopelessly lost, barely able to see the road in front of him. Today *The Runaway Bus* works better as a historical weather document than a comedy thriller: from it we can see that back then fog really was quite something.

In most cinema, fog is a malevolent, sinister entity. It's *the* malevolent entity in John Carpenter's *The Fog*, and the claustrophobic silence of the fog around the house in *The Others* serves only to heighten the tension that stretches the film as taut as Nicole Kidman's performance. What is therefore striking about the fog in *The Runaway Bus* is that there's nothing particularly extraordinary about it: the characters regard it as little more than an inconvenience worthy of a resigned sigh and a droop of the shoulders. They're not in awe of it, they're not scared of it, it's just a fact of everyday life. In the early fifties dense fog was commonplace, had been for decades and would remain so until the Clean Air Act of 1956 sought to take the smog out of the fog.

Fog rarely gets a good press. Few people throw open the curtains in the morning and exclaim, 'Oh great, it's foggy!' Monet was a fan of fog – 'Without fog,' he said, 'London would not be a beautiful city' – but poets seem to prefer mists, particularly when they're swirling around a bit of mellow fruitfulness. I find something wonderful in the mystery of fog. It's possibly the most enigmatic of all our weather: rain, snow and hail fall, the wind blows and the sun makes shadows, but the fog just hangs there. It doesn't do anything, it just *is*. If there is a breeze it can move like smoke, but a thick blanket of fog will hoodwink our senses. It removes the features and land-marks that define our perspective and depth of vision; there is no horizon on which to fix our bearings. We can't see any-thing except the ground around our feet: it's probably as close as we come to understanding what it must be like to be blind. (In the Great Smog of 1952 there were several accounts of blind people in London helping the sighted to get home: one man in Notting Hill regularly led people to their homes from the Tube station.) It disorientates us, forces us into ourselves, sometimes in fog the only thing we can hear is our own breathing, our exhalations clouding in the cold and becoming part of the fog.

Fog is usually seen as something sinister, probably because we don't know what might be lurking in it (Dante came upon Satan through a fog in the Ninth Circle of Hell in the *Inferno*), and for centuries it has been associated with dampness, illness and bad omen, but also I think we're nervous of fog because it's the one aspect of the weather that turns our attention into our-selves, forcing introspection. A walk in the fog, with no visual or aural stimuli, is one of the few times when we're truly alone with our thoughts, when there's nothing to distract us, nothing to look at, nothing to listen to, just our own breathing, footsteps and thinking. Yet while we can walk through the fog and can see it all around us, we can never touch it: it moves away as we approach, staying just out of arm's reach, half stepping aside to let us pass, half taunting us.

We leave no impression on the fog, no footprints and no shadows. We can't collect it in a cup or measure it in inches. It just sits there, a collection of minuscule water droplets so small it would take seven billion of them to fill one teaspoon with water. These droplets shimmer into view when the air temperature and dew point get close enough together to form what is essentially a stratus cloud that sits on the ground, a cool mass of air trapped beneath a band of warmer air. We're walking in cloud when we walk through the fog; the heavens have come down to the ground. Fog usually forms when conditions at night are cold, clear and calm, and the ground radiates the heat absorbed during the day. As this ground temperature decreases, it cools the air above it to the dew point, where water vapour condenses into liquid and forms what is known as radiation fog. Fog also forms when warm, damp air travels over a cold surface, condensing the moisture in the air to form advection fog. At or near the sea the water vapour forms around tiny specks of salt in humid air; these sea fogs are the only ones to leave a trace – a slightly grainy, sticky feeling on the skin and on clothing.

The cloying fog that clung to London in December 1952 left more than sticky skin and a salty taste on the tongue. Investigations in its immediate aftermath estimated that the Great Smog was responsible for the deaths of around four thousand people, while a 2004 investigation raised that figure threefold, with a hundred thousand more falling ill as a result of the smog. An anticyclone and a complete absence of wind had combined to leave dust and chemicals hanging over the city between 5 and 9 December. There'd been a particularly cold spell at the start of the month, which meant Londoners were burning more coal than usual to keep warm. Domestic coal back then was not of the highest quality – Britain needed to export the good stuff to keep the national coffers topped up – it was dusty and sulphurous, causing a yellow-grey smoke to vomit from domestic chimneys, along with that belched by power stations and trains. The coal smoke mixed with other chemical fumes from factories and found itself hemmed in by a temperature inversion, when instead of the temperature dropping the higher you went, the warmer air sat on top of colder air. This stopped the pollutants from escaping into the atmosphere and, with the stillness of the air, left a dirty yellow fog hanging over the city and its environs for four freezing, choking days.

London was used to heavy fogs and smogs. Centuries of coal-burning and then, in Victorian times, the growth of industry, had meant that 'pea-soupers' or 'London particulars' were part of the capital's folklore, as evidenced in the novels of Dickens (the opening of *Bleak House* contains the best evocation of fog in literary history), Robert Louis Stevenson and Arthur Conan Doyle. As far back as Edward I reservations had been expressed about the effects of so many people burning coal in such close proximity, something that had concerned Elizabeth I too. In 1661 John Evelyn suggested to Charles II and Parliament that something needed to be done about the 'catharrs, phthisicks, coughs and consumptions which rage more in this one city than in the whole earth besides', but nothing was ever done and the fogs got thicker, smellier and more dangerous. A smog of 1901 was so thick that people in the East End of London marvelled that they couldn't see their own feet, and my mother recalls coming within a single footstep of going face first into a pond while walking to school across Blackheath in a 1947 pea-souper.

The 1952 smog, however, was like nothing else. Football matches and greyhound meetings were cancelled, people groped their way along the streets, traffic was at a standstill and animals asphyxiated at Smithfield market. A performance of *La Traviata* at Sadler's Wells had to be abandoned before the interval as a large proportion of the audience couldn't see the stage for the sickly yellow haze. Hospitals were overwhelmed with people complaining of respiratory problems, and when funeral directors began to run short of coffins it became clear that this was no ordinary weather event.

People walked the streets with handkerchiefs over their noses and mouths; the police wore masks as they directed the snail's pace traffic as best they could. But it wasn't on the streets that the smog did its worst, it was in the freezing terraced homes of the old and infirm, where the noxious fog found its way in through draughty window frames and badly fitted doors, a poisonous sulphuric miasma that homed in on the inhabitants, inflaming their lungs and literally suffocating them in their beds and their armchairs. They'd survived the bombs of the war only to succumb breathlessly and slowly to a fog that hid a slow, wheezing death in its choking haze.

The Clean Air Act of 1956 saw that the London air would be gradually cleansed of its worse pollution excesses, although remarkably the Conservative government had reservations about its introduction. Harold Macmillan, Chancellor of the Exchequer at the time, twice ruled out legislation before the Act was passed on the grounds that 'an enormous number of broad economic considerations have to be taken into account'. However, it was soon realised that the cost of implementing smokeless zones was minuscule compared to the wider financial losses the smogs were causing, and the Bill finally went through Parliament. The effects were not instant – the Lewisham rail crash the following year, which killed eighty-seven people, happened in thick smog – but gradually improvements could be discerned and by the sixties London fogs were largely just that: fog, without the poisonous elements that turned it brown and yellow and caused polished silver to turn black on mantelpieces.

So when Frankie Howerd was climbing signposts and feeling his way along walls it was no mere hamming for the camera: to the cinemagoer of 1954 it was all perfectly plausible, memories fresh from the realities of a long December weekend that made them grateful the fog was just on the screen rather than hanging there in front of it.

When I was a boy, being allowed to stay up on New Year's Eve was quite a treat. Best of all was what happened at midnight. Once we'd watched the year being rung in by Andy Cameron and a bagpiper on the Sony Trinitron in the corner of the room, my sister and I would find ourselves being zipped into our coats and lifted into our wellingtons. Then we'd all go outside into the back garden, our parents would put their fingers to their lips and we'd listen. As our cheeks flushed red with the cold and our breath clouded in the night air we'd hear, carried up from the Thames, the sound of the ships on the river all sounding their foghorns to mark the beginning of the new year. It was magical; I hardly dared breathe for fear of missing any of it. It was a tradition that dated back beyond my sister and me: when my parents were first married and moved for a short while well away from the river, at midnight every New Year my grandmother would put on her coat, walk to a phone box, phone my mother and hold the receiver outside the door so she could hear the foghorns.

Maybe this is why I've grown to love the fog despite its malevolent history of agues and 'phthisicks'. For me, the foghorn is a happy, reassuring sound that symbolises newness and fresh starts rather than danger. As I write this the foghorns of Ireland have been switched off for a year now: when I first moved to the shore of Dublin Bay they were almost a reminder of home when they sounded on those regular foggy nights but now the horns all around the island have fallen silent, apparently for ever.

It was another man far from home who invented the foghorn. Robert Foulis was born in Glasgow in 1796. He studied surgery at Glasgow University for a while, but gave it up due to ill-health and became apprenticed to an engineer instead. At the age of just twenty-one he suffered the double blow of the death of his wife and child during childbirth, an experience so heartbreak-ingly awful that he fled halfway across the world to escape its memory. He had intended to make eventually for Ohio but a storm forced his ship to divert to Nova Scotia; he stayed in Halifax for a while, making a living as a portrait painter before landing the job of deputy land surveyor in Saint John, New Brunswick, which is where his life and career really took off. In 1825 he opened the province's first iron foundry, married again, had a daughter, became involved with some of Canada's earliest steam ships, opened a school of arts in the town and found the time to patent a gas illumination system designed for use in lighthouses. But his most enduring contribution to posterity came about purely by chance.

One night in 1853 he was walking home in fog when he heard his daughter practising the piano as he approached the house. Realising that he could hear only the low notes, he wondered whether he could turn this discovery into some kind of audible warning for ships in fog. He devised a steam-powered horn that produced a low-frequency sound and lobbied the Saint John Board of Trade to install it on Partridge Island in the city's harbour. For some reason Foulis's plans were given by the town leaders to another man, who set up the foghorn in 1859 and took the applause that was rightly due to Foulis. Although five years later he was at last credited with being the originator of the idea, Foulis never patented his foghorn. An American realised this, took out the patent himself and made a fortune. Foulis died penniless in 1866.

Bells and guns had usually been used as fog warnings but the manpower and cost involved were a constant headache for maritime authorities, especially as their effectiveness was far from consistent. An automated system, particularly one as clearly effective as Foulis's, was an obvious fillip for maritime safety and his invention soon spread around the world. The UK's first foghorn was established at St Abb's Head lighthouse in Scotland in 1876 and within a few years every lighthouse in Britain and Ireland was equipped with one. At the start of the twentieth century John Northey of Toronto invented the diaphone which, inspired by Wurlitzer organs, used compressed-air technology and this became the foghorn sound of the century across the world and the one we are most familiar with today.

Advances in navigational technology have meant that since 1995 fewer and fewer foghorns have sounded around our islands. The Scottish Lighthouse Board switched off their foghorns that year, concluding that they were no longer necessary in the new age of satellite position-finding aids, with Ireland following suit in 2011. A few horns still sound occasionally in England and Wales but the comforting throaty parp echoing around the darkness or through milky-white fog is largely a thing of the past, and the world is all the poorer for it.

Perhaps the horns' greatest epitaph is a recording made one night on the edge of the North Sea in 1982. A Dutch radio station staged an hour-long live 'foghorn concerto', or *Misthoornsconcert*, linking up transmissions from nine foghorns along the coasts of France, Germany, Belgium and the Netherlands. Strange as it may seem, it's an absolutely beautiful thing, wonderfully evocative of a dark night on the North Sea with a smoky sea fog drifting over the European coast respecting neither political boundaries nor coastlines. The horns range in sound from a deep, solar-plexus-hammering parp to a distant mournful cor anglais, cadences pitched at perfect and plagal, chords formed of notes played in different countries, an international concerto of weather, a musical ode to the sea and the protection of those on it from the elements. It's almost a musical version of the shipping forecast: if the latter is poetry then the minimalist, rhythmic tones of the North Sea foghorns provide a soundtrack.

As I played the recording, looking out at the lights of Dublin Bay, I sensed that I was coming to the end of my journey into the world of the weather. The Dutch foghorn concerto, sending its song of safety out across the sea, had brought me back to the shipping forecast and I realised I was going to have to make one last journey to bring the story to a conclusion, a pilgrimage to an appropriate place to gather together everything I'd learned. I knew there was only one destination, a truly historic outpost of the weather. I would head for the very frontline of our meteorology, a place where our weather first comes ashore, straight in off the Atlantic. It's a place whose name dates back to the days of Robert FitzRoy, a name that whispers of late nights by a softly playing radio. As a new year dawned unheralded by foghorns I headed west towards the sunset, the horizon and the Valentia weather station.

18

VALENTIA, WEST OR SOUTH-WEST FIVE, MODERATE OR OCCASIONALLY ROUGH, GOOD

For a craggy lump of wet rock on the furthest-flung fringe of Europe, Valentia has plenty of reason to tug at the sleeve of the rest of the world. The island, seven miles long and two wide, sits snugly against the western coast of County Kerry at the far south-western tip of Ireland, off the beaten track and on the face of it no more remarkable or unremarkable than any other island off Ireland's west coast.

Yet Valentia is very special indeed. The first working transatlantic cable came ashore here at Foilhommerum Bay in 1866, bringing Europe and America within virtually instant contact for the first time (the cable-laying had been carried out by the *Great Eastern*, the ship cooed over by the ill-fated folk aboard the *Royal Charter* seven years earlier). The island's slate quarry once produced what was believed to be the finest material in Europe, good enough for the Houses of Parliament and the National Gallery in London, and in 1992 a university student discovered the fossilised footprints of one of the very first land mammals ever to walk the earth near by; prints made by a tetra-pod some 385 million years ago and the oldest in the northern hemisphere.

Most notably for this story, Valentia is the name of arguably the most famous weather observation station anywhere in the UK and Ireland. Robert FitzRoy had selected Valentia to be among his original network of fifteen telegraph stations in 1860; indeed Valentia would have been one of the first names on his list. For one thing, the island already had a direct telegraph link to London in preparation for the transatlantic cable, and for another Valentia is situated in the very face of the weather, first in line for the arrival of every weather system that moves in from the Atlantic to pass across Ireland and Britain. Valentia has been a meteorological celebrity ever since and, as a fixture of the late-night extended shipping forecast, its fame is greatest among night owls, seafarers and the weather-savvy.

The morning after I arrived on the island I had breakfast looking out across the water at the slopes behind the fishing vil-lage of Portmagee, where a bridge links the island to the mainland. The location of the ridge I could see across my eggs and bacon means the sun hits the island later than most other places. It was a cold morning, and Ireland was shivering through a particularly cold few weeks. There'd even been snow on Valentia, a rare occurrence indeed, but now it was all gone save for the lines of foamy white wave tips undulating along the shaded slopes across the water.

In the afternoon I travelled to the eastern end of the island, to Knightstown, the main settlement and the harbour from where the summer ferry arrives and departs with the tourists. It was very quiet on this crisp January day; I was one of few people around to appreciate the sun shining on a bright blue sea, and the colours everywhere made bolder by the pureness of the light. I escaped the cold by ducking into the Royal Valentia Hotel, a grand Victorian construction dominating the harbour front, which once put up the future King Edward VII when he visited the Valentia observatory in 1869. It's a big, dark, rambling place, and even though a welcoming fire roared in the fireplace there was still the chill in the air and a faint whiff of bleach of a bar that had only just opened for the day.

With the place to myself I ordered lunch, took a seat by the fire, looked out at the sun sparkling on the water and thought about what had brought me here. It was at Knightstown that the first Valentia telegraph cable had come ashore in 1858. Two years later Robert FitzRoy had sent a set of his weather instruments to Mr R. J. Lecky, a member of the Royal Meteorological Society who ran the slate mine and also managed one of the telegraph companies operating from the town. A busy man who wouldn't have been able to commit to making the regular weather reports himself, Lecky coached a telegraph clerk by the name of E. O'Sullivan in how to use the instruments and it was the latter who, at eight o'clock on the morning of 8 October 1860, sent the first weather report from Valentia.

On FitzRoy's death in 1865 the Board of Trade decided that he had, in publishing his weather forecasts, far exceeded his brief at the Meteorological Department. A parliamentary committee declared that the science he was employing was not exact enough to justify so much expense, and six months later ordered that both the storm forecasts and weather forecasts be halted. Storm warnings would resume just over a year later, but the weather forecast did not return until 1879. This didn't mean the collection and collation of weather data were stopped altogether: the department just returned to the specific purpose of statistical analysis and collation for which it had been founded. Indeed, on the recommendation of the Royal Society the number of reporting stations was increased and self-recording instruments provided. In 1867 the Meteorological Department was released from the Board of Trade into the supervision of the Royal Society and renamed the Meteorological Office. Kew Observatory in south-west London was made the centre of operations and one of the first things the new Met Office did was fund a weather station at Valentia. A lease was taken on Revenue House on the southern shore of the island, the necessary conversions were made and the weather observatory was ready for business by the spring of April 1868.

The man appointed to run it was a former Royal Navy navigation lieutenant, the Reverend Thomas Kerr, fresh from training at the Kew Observatory. It must have been a daunting prospect for a man who'd become used to the bustle of London since giving up his life on the sea; after all, he'd just begun to settle in the city and had enjoyed both his training and the admission to metropolitan scientific circles that went with it. Valentia must have seemed like the end of the world. Nonetheless, in early June 1868 he hauled his trunk under the great arch at Euston Station and took a train to Liverpool, then a boat to Dublin before making the long rail journey across Ireland to Killarney. From there it was forty miles in an open horse and carriage to Cahirciveen, the nearest major town to Valentia, three miles by pony and trap to the shore and finally the small ferry boat over to the island itself. Being posted to Valentia was far from a snub, however: the importance of its location meant that the job came with some prestige, but the journey and the prospect of living on a small island off the coast of Kerry in an area blighted by poverty and still carrying memories of the devastating famine of the 1840s must have caused him some degree of apprehension. He arrived on the island in the middle of June and took delivery of the necessary instruments shortly afterwards. Having set up and calibrated them, Kerr was able to take the first readings of the Valentia Meteorological Observatory on 1 August 1868. He went on to run the operation entirely alone for six years until, with his health beginning to fail in the summer of 1874, a local man named Michael Sugrue was recruited to assist him. Kerr's health did not improve and the first official Valentia meteorologist died on the island in August 1875, at the age of fifty.

John Edward Cullum had been dispatched from Kew as soon as the grave condition of Thomas Kerr's health became clear. The appointment of twenty-five-year-old Cullum, who had been a magnetic assistant monitoring changes in the earth's magnetic fields at Kew, just a hop and a skip from his home in Richmond-upon-Thames, was originally intended to be a temporary one but once he'd started work on the day after Kerr's death he was to remain at the observatory for the next forty years. During Cullum's tenure Valentia stayed, despite its remote location, at the forefront of meteorological technology. For example, when an electrical anemometer arrived from Kew in 1888 and was set up on Ballymanagh Hill, one of the highest points on the island and in the firing line of the Atlantic winds. The anemometer was secured to a concrete base and linked to Revenue House via copper wires suspended between poles. It became an immediate local curiosity, not least because

it was visible from miles around. Local boys would amuse themselves by throwing stones at the sensitive apparatus; a problem
that became such a threat to its well-being that Cullum had to
ask the local parish priest to hector the culprits from the pulpit.
The results were immediate, but after only a few months of
service the instrument was dismantled and returned to London.
The concrete base is still there, a local landmark known to this
day as Cullum's Cup.

The major event during Cullum's stewardship, however, was
the move away from Revenue House and the island itself.
Evocative part of the shipping forecast it may be, but the
Valentia weather station has not actually been on Valentia for
more than a century. In March 1892 operations transferred to
the grander surroundings of Westwood House on the outskirts
of Cahirciveen, almost directly opposite the original site across
the water. The house had become available on the death of its
owner and, apparently aided by some vigorous lobbying for a
move to the mainland by the new Mrs Cullum, the whole
operation crossed the water to take up residence in the building which remains its home today.

In their more sumptuous surroundings, Cullum and Sugrue
soon settled into their routine of taking observations of wind,
rainfall, temperature and sunlight levels every two hours between
eight o'clock in the morning and six in the evening, then again
at ten o'clock, and communicating their results to London and
beyond. Their dedication was extraordinary: Cullum, for example, didn't miss a single day through ill health until May 1911,
when he was incapacitated by an attack of gout: an unblemished
attendance record of thirty-seven years. His retirement in 1915
brought the curtain down on the career of another of the great

unsung meteorological men of the Victorian age. He died three years later, in the aptly named Oxford district of Summertown.

After lunch I took the coast road along the south of the island to see if I could find the site of the old Revenue House. I knew that the building had been knocked down in 1939 and the stone used in the construction of a new church at nearby Chapeltown, but I was sure there had to be some remains or a marker to commemorate such an important part of Valentian and meteorological history. It took a few trips backwards and forwards but eventually I found it: an old length of low jagged wall like a set of broken teeth at the edge of the road, almost hidden by the bushes and trees growing on the other side. Close to the ground was an old plaque. '*Ar an láthair seo a bhí rhéadlann bhéal ínse ar dtús,*' it said, 'original site of Valentia Observatory 1868–1892'.

I walked through the gate and a gap in the wall that seemed to coincide with a set of French windows I'd seen in an old photograph at the National Library of Ireland back in Dublin. It was here that Thomas Kerr arrived after that epic journey from London and looked around his new surroundings, and it was here that he died on a summer's day in 1875. The view was still pretty much exactly as it would have been in his day, the Atlantic to the west, the view in that direction punctuated only by the steep-sided Skellig islands in the distant haze, and the Kerry mountains ahead and to the east. He would have noticed the sun arriving later than expected as it hauled itself up over the ridge across the water just as I had; a view he would have seen every day during his years here as a pioneer of the weather and, I hope, one that more than compensated for his transplantation in middle years from the great city of the Victorian age to a remote two-storey building with French windows on the furthest edge of the continent.

The next morning I headed for Cahirciveen, a typical small Irish town betraying now familiar signs of recession: bare, dusty shop fronts with optimistic 'To Let' signs propped in the windows. It's a popular tourist spot in summer, having been the home town of the 'great liberator' Daniel O'Connell, but in winter it's quiet and hunkered down, hibernating until the coaches and sunshine arrive. The activity of the weather station on the edge of town knows no such seasonal fluctuations; it just observes them, methodically undertaking the same daily routines as John Cullum and Michael Sugrue had done when they first moved to Westwood House more than a century earlier.

Cullum had been succeeded in 1915 by L. H. G. Dines, who came to Valentia from Eskdale in Cumbria with a name of some meteorological celebrity as his father, W. H. Dines, had developed a number of weather instruments including the Dines pressure tube anemometer and the Shaw-Dines microbarograph. Under L. H. G. Dines the station flourished: the staff increased from two to five, a library was established, an instrument workshop set up and Valentia became a popular testing ground for new equipment. The operation ran with barely an interruption even during the War of Independence of 1920–1, when the single reported disruption other than to services such as mail delivery was the theft of a theodolite and a field telephone by the Irish forces.

In 1922 Dines moved on and a Mr C. D. Stewart came to take over, only to walk straight into the maelstrom of the post-independence Irish Civil War. Stewart had barely been in Ireland for two months when he reported in August how 'all land communications with this place broke down from the 5th August, the Republican forces wrecking the railway and the telegraphic wires. On the 23rd the Irish Free State forces took the town of Cahirciveen after some fighting, most of the actual shooting taking place in the vicinity of the Observatory. The whole operation was easily visible from the Observatory windows. The 1800 and 2100 observations were incidentally rendered extremely unpleasant by the constant crossfire of the two sides.'

Stewart undersells the dedication of himself and his team here. The fact that the two evening observations were made at all is remarkable: the observatory staff had crept out among the flying bullets carrying white flags. There had been no question of even postponing the readings, let alone missing them altogether.

While the readings continued to be taken during the hostilities, communicating them to London when the telegraph and telephone wires were repeatedly cut was almost impossible. The Valentia team was to all intents and purposes cut off from the rest of the world, sometimes for weeks at a time, but when mail and wires eventually got through to Cahirciveen they would include stentorian complaints from London at the non- or late appearance of data. A letter in the observatory files shows Stewart at his restrained best in the face of such laughably intransigent bureaucracy. In a reply to one such upbraiding from Kew he writes, 'It does not appear to be understood at Headquarters that the isolation is no fault of the staff and cannot be remedied by us. Complaints of delay in receipt of returns etc are not merely unreasonable but, in the circumstances, are trivial. Wires calling for the urgent rendering of returns are ludicrous, since to commence with, the wire never reaches us in less than a week and more frequently takes three weeks.'

Curiously, the observatory stayed under British jurisdiction until the mid-thirties, when the Irish Free State Meteorological Office assumed control, but the staff's tremendous longevity of service continued: Michael Sugrue, who'd become Thomas Kerr's first assistant when he joined the station in 1874, retired in 1926 and his replacement Michael O'Shea stayed in his post until 1977, meaning that in over a century the job had been done by just two people.

I drove up the sweeping driveway to Westwood House, past a square, modern building and a number of strange-looking instruments scattered around the grounds. In the sunshine, the yellow-painted house looked as spruce as it must have done when it was first built. I'd come to see Keith Lambkin, the Chief Scientist, who had offered to show me around and explain the routines and research carried out by him and his team. Being an idiot, I somehow failed to find the main entrance, circled the building and found myself tugging at a locked side door. Someone heard my frantic handle-rattling, unlocked the door and let me in, kindly refraining from asking how I'd managed to miss the front door altogether, despite parking in clear view of it, and escorted me to an office whose door bore the legend 'Chief Scientist' in gold leaf on a block of varnished mahogany.

I'd expected Keith to be some kind of intense egghead with the sort of humourless demeanour that surely went with the monotonous routine of collating statistics in an old building at the far end of a town at the far end of a country at the far end of a continent, but instead a tall, smiling, dark-haired young man bounded across the carpet from behind a large oak desk piled with papers, shook my hand, ushered me in, pulled out a chair for me to sit on and disappeared in search of coffee. His office is at the corner of the building, a large, high-ceilinged room that must have originally been the drawing room. Huge windows looked out towards Valentia, the Atlantic and the mountains, the very windows through which the staff must have observed the shootout before deciding whose turn it was to grasp the white flag and tiptoe to the rain gauges.

Westwood House is a lovely old building bathed in the region's clean natural light that falls through the windows and on to bookcases, portraits of stern-looking Victorians, rugs and wide wooden stairs. The skirting boards are scuffed but the paintwork is clean and fresh, sure signs of a place that's as well loved as it's well used. Voices and laughter echoed from all over a building that quite possibly still resonated with C. D. Stewart's exclamations on receiving another of those farcical bureaucratic bollockings from across the Irish Sea.

Keith burst back into the room bearing coffee and biscuits, and so began an afternoon spent with one of the most enthusiastic people I've ever met. If ever someone was suited to a job – and vice versa – it's Keith Lambkin, a man of clearly massive intelligence who's lucky to be doing a job he loves, and knows it. He has a rare gift for explaining difficult concepts to whiskery simpletons wearing vacant smiles and waving Dictaphones around too. Keith made his work come alive with clear, simple, passionately expressed explanations. By the time I left, I'd been so infected by his relentless enthusiasm I'd practically vowed to go off and become a chief scientist myself, even if I did have only a failed physics O level to my scientific name.

Originally from south County Dublin, Keith had been sent to Valentia, despite having expected a posting in the capital.

'It was a massive culture shock, huge,' he said, anticipating my next question. 'Up there you're used to your range of local pubs and your nightclubs and what have you, down here Cahirciveen has the one nightclub for about a twenty-mile radius so you get busloads coming in on a Saturday night. It's massively different. I got married last year and my wife moved down from Dublin, so I'm now watching her go through the same culture shock I went through when I first arrived.

'It has taken some getting used to all right. My sister knows the area well – she used to come here a lot on holiday – and she said to me not long after I'd moved here, "I bet if I sent you a postcard just addressed to Keith Lambkin, Cahirciveen, it would get to you." I said, "No chance", we had ten euro on it, and after three days I phoned her and said, "It's not here, you owe me ten euro." The next morning I came into work and there's the postcard sitting on my desk. I asked around and it turned out that the local maintenance man who does some work here is one of the local firemen, and he was talking to the other local fireman who's also the postman. He'd got this post-card and didn't recognise the name, but then they thought, there's a new guy in town working at the observatory, it must be him. And that's how it arrived on my desk.'

I commented on the history of long service at Valentia, joking that I'd see him back here in forty years. He laughed, but from the way he glanced sideways out of the window towards the sun-dappled sea as he did so I could tell that it really didn't seem like such a bad idea.

From there we got to talking about the dedication of the men who'd continued making the observations during the Civil War.

'That was literally just out here,' he said, waving a hand towards the window, 'but the golden rule is you have to get your observations done, without fail, no matter what. But even so, that showed dedication that's second to none.'

He told me how the staff today are still doing exactly the same thing, recording the weather both visually and from instruments and sending the data off to Dublin where it's combined with information from other stations to produce the weather forecasts.

'The place does have a fascinating history all right. I'm here three and half years and I'm still learning; some of the old-timers who've been here years, they know the place inside out. Apparently people used to live here, there were numerous people born in rooms all over the house over the years – it's fascinating stuff.'

I'd arrived with an impression of Keith spending his time ambling out to a Stevenson screen a couple of times a day, squinting at the instruments inside, scribbling a couple of things in a notebook and ambling back in again to stare out of the window for a while and do a bit of Sudoku. I was completely wrong. He's a busy man.

Valentia is in an important place because the prevailing winds from the Atlantic come in from the west; the observatory is one of the first places to sample these air masses as they arrive before they pass over Europe. Valentia provides one of the first visual checks in Europe on wind direction and the directions in which the weather is moving, and these days it's also very important for pollution samples.

'We measure air and rain pollution here at Valentia, something that's done in exactly the same way at different spots around Europe,' said Keith. 'There's always a background count of pollution anyway, so the relatively clean air coming in over the west of Ireland contains a steady background level of pollution, which the rest of Europe can use as the basis to come up with a true reading of pollution levels over the continent.'

Valentia is also one of a select network of global weather stations to launch weather balloons, which is something they've done since 1943. Modern weather forecasts are created on three-dimensional models: there are various observations taken on the surface of the planet, but a vertical profile is needed too. This is where the weather balloons come in.

'They're interesting because there are nearly a thousand of them all launched in unison at exactly the same time all around the world, twice a day, sometimes four times a day,' said Keith. He looked at his watch. 'Actually, we'll go and watch a balloon launch in a bit,' he promised, to my schoolboyish delight.

He went on to explain that Valentia is the only place in the country that manages solar radiation levels and stratospheric ozone. There's the ground-level ozone, the kind that's bad for us, but the high-level ozone in the atmosphere creates a protective shield for the planet. Valentia measures the strength of that ozone.

'Another interesting thing we do is phenology, the study of plant life cycles,' Keith went on. 'We have these specially cloned trees in the grounds that were spread around Europe in the early sixties and we take measurements from them. The beauty of it is that, being cloned, all these trees across Europe are biologically identical. The only differences are in the soil and the climate around them, making them an excellent climate–change indicator.'

Valentia is part of the International Phenology Group, and today there are more than fifty such gardens across Europe, all comprising biologically identical trees grown in Hamburg in the late fifties. Most of the gardens are clustered around the centre of the continent; Valentia is the westernmost point of the network, making it a key player in spotting and monitoring the changes in the seasonal cycle of the trees in their care.

We finished our coffee and set off out into the grounds. Keith had suggested we meander slowly towards the modern building near the road which houses the weather balloon launching system and on the way he'd show me around some of the instruments and work sheds. I agreed on the grounds that any plan involving the words 'meander slowly' is all right with me.

'This is the international standard way of measuring rainwater,' he said, indicating what appeared to me to be a funnel and a bottle, behind which I assumed there was a complicated scientific method. 'Essentially, it's a funnel and a bottle: there's no complicated scientific method behind it at all. The rain falls out of the sky, into the funnel and into the bottle, and from this we measure the pollutants in the atmosphere. It's a very good way of sampling what's in the atmosphere – you know, in case there's another Sellafield or whatever.'

Further on we came to a large white box mounted on a pole.

'This is a ceilometer,' explained Keith, 'which acts a bit like the Bat-Signal and measures the height of the clouds overhead. It shines a strong beam of light straight up at the clouds, a detector up on the mountains finds where the beam hits the cloud and measures the angle, then, through simple trigonometry, it is pretty easy to work out the height of the cloud base.

'This antenna is interesting,' he said as we approached what looked to me like a fairly run-of-the-mill one. 'It's run by the British Met Office and it's a lightning-detection antenna, basically a very fancy radio receiver.'

When a bolt of lightning strikes the ground it generates a radio shockwave, which travels immense distances through the atmosphere. Each of these shockwaves gives out a specific signal that is picked up by antennae such as the one I was admiring and, when coordinated with other antennae around the world, allows us to pinpoint exactly where the lightning struck. So, for example, if the Valentia antenna picked up a particular lightning signature it would arrive in France at a slightly different time. Ditto in, say, Lerwick: they'd also pick up the same signature, again at a slightly different time.

'This one is so accurate and so sensitive it can detect lightning strikes as far away as Canada. It covers huge distances and is great for tracking storms, something that's very useful to the aviation industry. In fact, it actually works too well – there are so many lightning strikes around the world that they have to set a threshold so it only picks up waves within a certain set of parameters, otherwise it would just be going all the time.' I thought back to Roy Sullivan. If he'd been around today it would be possible to track his movements simply by monitoring the occasions when he was struck by lightning, like a deadly GPS.

We walked across the site to what appeared to be a large shed sunk into the ground. A pathway down the side of the shed led us to a doorway nearly at ground level. Inside, one wall consisted almost entirely of the very bedrock of the land itself. A sleek white machine hummed near by.

'This is our seismic vault, half built into the bedrock,' said Keith, his voice echoing around the space. 'In the late fifties the US Geological Survey contacted Met Éireann because they were interested in setting up a network of seismic stations around the world. "We'll come over and put it in for free," they said, "on the condition you send us back the traces once a week." Back then the seismographs were like the ones on TV, the big drum and paper things with the needle going backwards and forwards, an electrical current burning a trace on electrical paper. Nowadays they're all electronic but work on the same principle: a pen sitting on a spring that's attached to a chassis that's sitting on the bedrock and picking up the tiniest vibrations running through the ground.'

The equipment is incredibly sensitive, so much so that it picks up the movement of the waves on the Cahirciveen coastline. When an earthquake occurs the shockwaves pass right through the centre of the earth as well as causing chaos on the surface. The waves pass through the earth's mantle, the outer core and the inner core and then out the other side. As it passes through these different levels each shockwave refracts, all information picked up by the little needle half-submerged here in the south-west of Ireland.

'This guy here picked up the tsunami and the Japanese earthquake, things that happened right over on the other side of the world,' said Keith with a tinge of pride and a hint of awe. It's an extraordinary thought as we watch the devastating scenes from, say, Japan or Christchurch, that the force created by those quakes actually passed beneath our feet.

It seemed pretty generous of the Americans to go around the world dishing out expensive equipment to obliging meteorologists, but of course there was an ulterior motive.

'What they were doing was building a network of seismic stations to keep an eye on Russian nuclear testing,' confirmed Keith. 'Atomic tests have similar footprints to earthquakes, so we've pretty much got the Cold War to thank for this.'

I thought of the sunny vista outside, a small town in the far south-west of Ireland going about its business, not suspecting that for nearly thirty years it had been part of the front line of the Cold War. It also reminded me that the only proper job Stalin ever had was the two years he spent as a young man working as a meteorologist at the weather observatory in Tiflis (now the Georgian capital Tbilisi) before his revolutionary activities became a full-time occupation. It seemed extraordinary that Keith and Uncle Joe had done the same job. I couldn't imagine Keith as a ruthless dictator, but there was certainly something pleasing about the thought of Stalin the weatherman. We left the seismic vault and walked towards something that, at last, I recognised: the white slatted square box on legs of the Stevenson screen. We'd had one of these at school, but such was my ignorance then that I'd thought it was a beehive and would spend less engaging lessons staring out of the window on the lookout for the bees. Stevenson, incidentally, was Thomas Stevenson, famous as the head of the engineering family who built most of the lighthouses that guard the coast of Scotland, more than thirty of them in fact. Despite this incredible record in engineering and meteorology, Stevenson is probably best remembered as the father of the novelist and fan of Robinson Crusoe's brolly, Robert Louis Stevenson.

'This is your classic weather station, your Stevenson screen, in which you have your wet-bulb and dry-bulb thermometers,' said Keith, opening the box to show me the thermometers inside. 'We also take ground-level temperatures and soil temperatures here, which is important for agriculture. Now, generally we never see heavy frost or snow at Valentia as we have a relatively warm breeze passing through from the Atlantic. But when we had that very cold weather recently the moisture around the wet-bulb thermometer actually froze solid.'

There is a way of still measuring the temperature when the wet bulb freezes, but the observer needs a special set of tables with dedicated coefficients. As the Valentia wet bulb had never frozen before, at least not in living memory, nobody knew where that set of tables was. The Valentia observers turned Westwood House upside down in searching for them but to no avail. Frantic calls were made to weather stations around the country until luckily a copy of the relevant tables was faxed to Valentia just in time to meet the deadline for taking the readings.

Keith laughed nervously. 'Imagine, after those lads crawling on their stomachs with their white flags with bullets flying around still getting the readings and we end up missing one because we couldn't find the right sheet of paper ... '

It was a cold day but the shudder that went through Keith at that moment had nothing to do with the temperature.

We moved across to two white plastic funnels on poles. 'Rain gauges. We have the old manual one and also these electronic ones that have a kind of see-saw effect that means as the water falls it tips into the container. But you still want to keep your old system so a) your records are consistent with the historical ones having been taken the same way and b) you're not just relying on a single source. It's safer to double up.'

We then moved to a platform containing four instruments set out in a line that seemed to progress in technology from Jules Verne to Douglas Adams.

'This is the solar radiation side of things,' said Keith. 'I like this because you know that drawing of the ascent of man? These instruments give us the same sense of evolution. This one here,' he said, indicating what appeared to be an end-of-pier fortune-teller's crystal ball sitting on a cardboard bookmark, 'is a Victorian instrument called the Campbell–Stokes sunshine recorder. It works on the same principle as when you were a kid and you'd fry ants with a magnifying glass.'

The glass ball focuses the sun's energy on to the card, burning the rays on to it to provide a record of the sun's strength as the day progresses: the deeper and more pronounced the burn, the stronger the sun's radiation. The card is removed at the end of each day and from the nature of the burn it's possible to deduce the number of daily sunshine hours. The Campbell–Stokes sunshine recorder has been used at Valentia for more than a hundred years, providing a valuable and consistent set of records, but the problem is that depending on the moisture content in the air it can leave a different impression on different days.

'The sun can be at exactly the same height and the same strength on two different days,' Keith explained, 'but if it's been raining, say, the card doesn't burn as much as it would on a dry day because of the dampness in the air.'

We moved on to the next instrument in the line, a low white plastic dish.

'This is a global pyronometer. It measures energy from the sun but also the energy bouncing off the clouds and off the ground; basically all the solar energy that the human face would feel.'

The pyronometer converts solar energy into voltage, enabling the meteorologists to take a much more accurate reading of the sun's energy than the Campbell–Stokes. A similar machine was next in the line; it had the same sensor but with a disc placed between the sun and the component taking the reading. This disc, Keith told me, 'protects the direct component from direct sunlight as it moves across the sky and allows us to measure the diffuse solar energy – what we call the back scatter'.

'This one here,' he said, standing next to a spindly, space-age machine that looked like it could be a chirpy robot with a heart of gold from a Pixar movie, 'is a solar tracker, a very expensive piece of kit with a simple job. He wakes up in the morning, follows the sun around the sky and then resets himself at night. The beauty of that is that anything attached to him points directly at the sun all day too, so these here are pyroheliometers, which record energy coming directly from the sun. So, what you should find is that if you take the reading from the global machine and take away the reading from the diffuse recorder you should get the same reading as the pyroheliometer.'

All this talk of sunlight made me realise just how cold it had become. We'd been outside for a good while, but Keith's passion for his subject was so absorbing I'd barely noticed the chill in the wind. It was time for the highlight of the day, however: the balloon launch, and as Keith led me towards the building where it would happen he explained the process.

'Fifteen years ago we introduced a unique system for launching the balloons. Before then, we'd do it manually from that little hangar over there, inflating the balloon inside, wheeling it out and letting go, which could be very dodgy in high winds. But now we've got it housed in that building there.'

The balloons are launched from the roof of the building, which looked like a cross between an airport control tower and a sewage pumping station. Inside a small cabin on the roof is a wide-mouthed cannon device running on hydraulics with a semi-automatic launching system. When the balloon is ready to go the cannon rises up, points in the direction of the wind and fires the balloon out, the instruments dangling beneath it.

'I still find it amazing to think that all around the world now there are hundreds of these balloons all getting ready to launch together. In fact there was an arts cooperative in New York that put together a project about when science and politics meet a couple of years ago, and they chose the weather balloons, all launching at the same time whatever the location, whatever the politics, to show an occasion when the whole world is united at one moment. They put together this multi-screen HD production of the world coming together for the launch of these weather balloons and it was premiered at the UN, shown to all the world leaders.'

Inside the building we climbed to the first floor, which is effectively mission control for the whole balloon operation. Looking around at the screens and flashing lights I realised this was starting to feel less like a tour of a historic weather station and more like a spin around the lair of a Bond villain. When Keith produced an example of the tiny panel that hangs from the bottom of the balloon, recording and transmitting vital data all the way up to the frontiers of space before falling back to earth, I was reassured that, in Bond terms, he was more Q than Blofeld.

'It's very light – it has to be in case it hits an aircraft or something as it drops after the low pressure bursts the balloon,' he explained. 'It's made up of a temperature sensor, a humidity sensor and, inside, a pressure sensor. You also have a location sensor to help us pick up wind speed and direction. These things go up incredibly high, up to thirty-five kilometres, which is something like three times higher than commercial jets fly.'

We climbed a steep staircase. Keith opened a door and led me out on to the roof, where in front of us was what looked like a metal shipping container, painted gunmetal grey. There were two small domes on top: the receivers for the information constantly being fired back to earth by the balloon. We were barely a couple of minutes away from launch and I was getting excited. I thought back to the well-groomed, shiny-faced kid in the pictures I'd seen in the Ladybird book that had helped prompt my journey of meteorological discovery months earlier. I bet he'd never got to watch a weather balloon go up. Tough luck, sucker.

'You'll notice that when the balloon comes out of the cannon it'll look quite flabby, not taut like a normal balloon,' Keith told me. 'The reason for that is the air pressure way up there drops dramatically, making the gas in the balloon swell. If we filled it right up then it would burst too quickly. Right, get ready, here we go.'

There was a whirring noise and a funnel around six feet across rose from the roof in a slow, elegant way that deserved to have a man in evening dress playing a Wurlitzer organ on top of it. The cannon's lid sprang back, the funnel tilted towards the direction of the wind and a white latex balloon emerged, slowly at first and then, when it was free of the cannon, racing for the heavens as if it had just realised its freedom and couldn't quite believe it. We stood in silence, heads tilted back, and watched as it shrank to a speck and disappeared into a clear sky of approaching dusk. The balloon would take roughly an hour and a half to reach its maximum height, right on the very edge of space, as high in the atmosphere as it's possible to go without leaving it. When it reached its zenith Keith and I would be enjoying a cup of tea in the Westwood House kitchen, still talking about the weather as darkness fell outside.

One thing I particularly liked about Keith was that he is as happy looking back as he is forward: for him the past is as important as the future, especially regarding Valentia. As a scientist he's obviously excited by new developments in technology and equipment that aid the pushing of the boundaries of existing knowledge as far as they'll go. Yet he's also in awe of the history of Valentia and unquestionably in love with the place. When I told him that I'd managed to find Thomas Kerr's grave in a small Church of Ireland cemetery just outside Knightstown he became animated with excitement, telling me he hadn't realised it was there and asking me exactly where it was. I knew it wouldn't be long before he would visit. But it was probably the story of the first ever Valentia weather report that impressed me most about Keith.

The year 2010 was the 150th anniversary of the first observation made at Valentia. A series of events was held, lectures and open days, local people having a rare opportunity to see what actually went on behind the big gates at the far end of the town. But while Keith could put his hands on the first logbooks from

Revenue House, which dated to 1868, no one seemed to know where to find the original record of that first observation, the one taken and transmitted on 8 October 1860 by the Knightstowntelegraph operator, who is remembered to history simply as E. O'Sullivan. It would, thought Keith, be a little odd to mark the sesquicentennial of that observation without having anything to show of it whatsoever.

So he set to work. The observatory archives threw up nothing; contacting University College Dublin, Met Éireann and the Meteorological Office in the UK drew blanks too. Weeks turned into months and Keith was no nearer finding that first Valentia weather observation. Then, at a conference in Cork, a delegate quoted an old newspaper report in a seminar and a light went on in Keith's head. At the same time a retired meteorologist whom he had contacted for advice got back to him with a newspaper article he'd spotted, which mentioned how Robert FitzRoy had placed his weather forecasts in *The Times*, beginning a month before that first Valentia observation. Delving into the *Times* archives, Keith eventually found the edition for 9 October 1860 and at once his quest was over. There in the table of the previous day's observations, slotted in between Liverpool and Queenstown (now Cobh, County Cork), was Valentia, the first time the name had appeared anywhere in connection with the weather.

What E. O'Sullivan had recorded – updated to equate to modern readings – was that the wind was west–north-westerly, one to three knots, the day was overcast, 11.7°C, there was no rain and there'd been a barometer reading of 1027.4. Pretty useless information on its own, but to a man as sensitive to Valentia's weather history as Keith, it was the holy grail.

At eight o'clock on the morning of 8 October 2010 Keith stood beside Valentia's Meteorological Officer, Feilim Hanniffy, as he took and transmitted the morning's weather observations one hundred and fifty years to the minute after a man named O'Sullivan had done the very same thing for the very first time. Keith's determination to find the record and the sense of ceremony he recognised in the anniversary make him one of the heroes of this book for me. As I headed back to the island through the dark I reflected on how there was a direct line from Keith back to FitzRoy himself, that same zeal, that sense of wonder at and respect for the elements and the same desire to understand and interpret them.

Later I crossed the bridge from Portmagee back to Valentia, after dinner and a couple of pints. It was a clear night and with minimal light pollution I was treated to the most incredible canopy of stars. Up there somewhere was the latest Valentia weather balloon, transmitting readings back to the grounds of Westwood House and propagating the work started here by Thomas Kerr, Michael Sugrue and John Edward Cullum. Up there in the same night sky beneath which they would have taken their readings and recorded them with meticulous accuracy in copperplate handwriting, a tiny panel of electronic circuitry was broadcasting the weather from the very edge of space down to a small, remote, historic and heroic building on the furthest fringe of a continent.

19

TIME AND THE WEATHER

The next morning I stood on the very edge of the cliffs at the westernmost tip of Valentia Island with my arms outstretched, the sun on my back, the wind booming in my ears and the sparkling Atlantic heading off to the horizon beneath a cloudless blue sky. And while I was there I thought about William Merle.

'In January there was warmth with moderate dryness and in the previous winter there had not been any considerable cold or humidity, but more dryness and warmth. In February, during the first week there was moderate frost, and after an interval of three days there was slight frost for another week.'

As descriptions of weather go it isn't one that's filled with vim and vigour. It isn't one that's filled with anything especially interesting at all, in fact. Yet in terms of our relationship with the weather these are very significant lines indeed, not so much because of what they say but when they were said and by whom. What sounds like some particularly forgettable small talk is actually the earliest weather record we have. It concerns the weather in the village of Driby in Lincolnshire, was written by Merle, the rector of that parish, and is a record of the weather there for the months of January and February 1337.

We know absolutely nothing about Merle other than his religious role at Driby and his apparent fellowship at Merton College, Oxford, although there seems to be no mention of him in any records there. His journal follows his movements between the two locations: some reports are of Driby, others concern Oxford. You might expect that the weather records are part of a more general diary or personal journal but no, Merle's journal really is just all about the weather. Despite writing diligently for seven years, almost to the day, he reveals absolutely nothing about himself. There are no personal anecdotes, no clues as to his character other than the meticulous maintenance of his weather diary suggesting an endearing fourteenth-century nerdishness; no mention of events in his life, no allusions to his religious calling and no noting of local events. There is one out-of-the-ordinary report in the journal but it's one that Merle must have included because he considered it to be weather-related. On 28 March 1343 he noted the following:

> Stormy with a very strong wind from the north-west, and with hail, rain, snow very often in the day. At mid-day there was an earthquake, which was so great that in certain parts of Lyndesay the stones in the chimneys fell down after shaking in very great agitation, and it lasted long enough for the *salutatio angelica* to be clearly said. The aforementioned earthquake was not felt at Oxford.

Other than that, it is literally just a record – sometimes monthly, sometimes weekly, sometimes daily – of the weather above and around William Merle. There is no hint as to why he started the journal or why he stopped. He just . . . did.

There is nothing going for it as a piece of writing (it's in Latin, for one thing), is of little more than passing interest to meteorologists and historians and seems to stop as suddenly as it starts (the last entry gives the date as 10 January 1344 but there are no notes to go with it). Whether the political and social upheavals of the time impacted upon daily life in Driby we'll never know (although news of them probably reached Oxford), but even if they did they were of no concern to Merle's weather journal, written in a careful neat hand in a brown ink on vellum. To the modern eye the *Considerationes Temperiei Pro Septem Annis* is utterly tedious, pointless and of no interest to anyone except William Merle. And I think it's fantastic.

I'm not entirely sure why I love Merle's journal so much. Maybe it's because much of the history we read, certainly of that period, is about sensational events featuring exceptional people, yet here's someone fairly ordinary, someone otherwise forgettable, someone far removed from the epicentre of great events who saw fit to record the mundane and the commonplace for reasons known only to himself. There is something very human about Merle's journal. We've no idea what it was like being Edward III sitting in council with his advisers discussing the budget available for a war with France, but we can certainly imagine a man getting up in the morning, noticing the chill in the air, looking out of the window and seeing a crisp, white sugaring of frost on the ground because we've all done it, we all know how that feels. We might not have kept a record of it, but for me Merle's journal brings the fourteenth century to life in a way that even the most accurately observed, CGI-drenched blockbusting account of the Hundred Years War wouldn't.

In writing his records Merle becomes one of us. We may snigger at the geekiness of a grown man keeping a carefully annotated record of the weather in his locality, but to this day the British male still displays a penchant for the nerdy, whether it be making Spotify playlists or compiling his top ten FA Cup Final goals. William Merle is in all of us in a way that other figures from history are not. Take Merle's account of the earthquake in 1343 – his addendum that 'the aforementioned earthquake was not felt at Oxford' paints a picture that goes beyond those simple words. You can imagine him travelling to Oxford, bustling into Merton College and asking his colleagues if they'd felt the earthquake and being met with blank looks, telling them all about how the chimneys crumbled and the tremors went on long enough for people to recite the *salutatio angelica* and his colleagues saying to him, 'No, we didn't feel a thing here, but tell us more.' We all enjoy telling a good story and from that entry I like to think Merle revelled in recounting that one. If his prose style is anything to go by he wouldn't have told it very well, but he'd have got the point across and enjoyed being the bearer of the news.

Otherwise it's the sheer relentless mundanity of the thing that makes it special for me. The first month, January 1337, saw the death in Florence of Giotto, yet at the same time, hundreds of miles away in an obscure corner of England, the weather was a bit warm and dry for the time of year. Somehow that gives the whole thing context, balancing the significant and the quotidian; it brightens the dark corners of history and brings January

1337 alive for the twenty-first century. There's such an otherness about the long ago that it's almost comforting to know that
people like William Merle were noticing the things we notice
and talking about the weather.

Who knows why he did it? Vellum and ink were presumably
not cheap in those days, so it was quite a commitment to keep
a journal of any kind. Yet he was consistent and meticulous for
seven years, until stopping suddenly that January day three years
before his death. Maybe, as the weather would have been seen
as the work of God, it was intended as a record of His works
in the sky? Maybe it was some kind of demonstration of
Merle's faith? Was he trying to see some pattern in the weather,
beyond the obvious changes of the seasons? It is likely he was
the only literate person in the vicinity, so was he keeping a
written record for the benefit of local farmers, for whom the
weather was vital to their livelihood? Thirty years earlier there
had been a dreadful famine across Europe when the harvest
more or less failed for three summers in succession, from 1315
to 1317: was Merle keeping records in order to spot trends and
patterns that might help with the planting and sowing in the
local fields? Whatever the reason, it seems unlikely he'd have
kept his records over such a lengthy period of time just on a
whim.

We have no idea how old William Merle was when he kept
his journal but it's written in what appears to be a neat and
steady hand to its very end; he didn't seem to be elderly and
there does not seem to be any gradual decline through illness.
The account just stops as suddenly as it started and it seems odd
that he had inked in that final date and not completed the
entry. Maybe he was keeping the record for the benefit of

someone else and they died or left the area, although this doesn't explain why he noted the weather on his visits to Oxford.

This sudden ending makes William Merle's journals even more tantalising. We know he lived for another three years, it's too early for the arrival of the Black Death and it's fairly safe to say that neither the Hundred Years War nor the Crusades ever went crashing through the villages of eastern Lincolnshire, but something made him start and something made him stop. I might be a bit of an old romantic here, but I like to think that maybe he simply saw the beauty in weather. The purple-orange pinks in the winter sunrises, the magical wonderland of a morning frost, the perfectly white fluffy clouds against a deep blue sky, the gusty tufts of billowing snow, the musky freshness of autumn mists, even the glory in the rampaging malevolence of the heavens darkening and unleashing a thunderstorm. William Merle was no poet, author or artist. Intelligent and literate, his Latin was his tool but in it he lacked the descriptive powers that could do justice to what he was seeing in the sky.

Merle made no contribution to furthering our understanding of the weather. He invented no instruments and left no theories. Others before and after would make much more significant contributions to our understanding, but he illustrates the human aspect of our relationship with the weather. He was only a Lincolnshire clergyman who emerges briefly from the mists of history to provide us with a record of parochial weather conditions that is of no use at all other than as a fourteenth-century curiosity, yet there was something in William Merle that appeals to me deeply. Maybe it was a sense of being unified with him in time. Like Merle I'd become fascinated by the weather and was looking to the skies with wonder and a desire to understand our relationship with it. I had books, libraries, the internet and the Keith Lambkins of the world to explain it to me; Merle just had the evidence of his eyes and his faith in his God. We both saw exactly the same things, but all I had over William Merle was the advantage of time.

The weather and time go hand in hand. Indeed, in many languages the words for 'weather' and 'time' are the same: *le temps* in French and *timpul* in Romanian, for example. Before we had clocks time was measured by the weather, the turning of the seasons being the surest sign of time passing. Some languages, when asking the time, ask what the hour is. The word comes from the Horae, the Greek goddesses of the seasons.

The weather marks time on a large scale too. During the Roman era our weather was much warmer to the extent that wine was made across Britain. Around the turn of the fourteenth century, not long before William Merle began his journal, the weather started to grow colder and North Sea gales battered the east coast. The fourteenth century brought the Black Death, the spread of which many historians believe started as the result of a massive flood in China that drowned up to seven million people and forced the black rat to migrate westwards. Around the middle of the sixteenth century began what meteorologists call the Little Ice Age, a three-hundred-year cold spell that started with the kind of vicious storm that scattered the Spanish Armada and led to persistently cold winters and devastating famines, most notably in Scotland in the mid-sixteenth and early-seventeenth centuries and towards the end of the seventeenth, which led to mass population migration.

The Thames froze regularly and the frost fairs on its icebound surface became almost an annual occurrence. The sea around southern England froze to a distance of up to three miles from shore in the harsh winter of 1683 and for an estimated twenty-five miles from the coast of Holland, paralysing ports and causing economic chaos. There was the 'forgotten famine' in Ireland caused by the winter of 1740–1, and in 1784 the weather grew so eye-poppingly cold that the Danish government seriously considered evacuating the entire population of Iceland.

Around 1850 things began to warm up again, leading to the prevailing weather conditions we enjoy today. In weather terms, however, time seems to be speeding up. The future of the weather is uncertain as climate change continues apace, while the possible economic consequences play far too great a role in our responses to it, just as they obstructed the Clean Air Act. But whatever the future holds there will still be a continuous line of people documenting, thinking about, exploring and predicting the weather, a line that starts with Aristotle and runs through William Merle, René Descartes, Francis Beaufort, Luke Howard, Robert FitzRoy, John Edward Cullum and Keith Lambkin.

As I stood on that cliff top and let the wind fly through my hair and make my jeans flutter around my calves, prodded and explored by it as probably the first person in Europe that this wind had encountered, I thought about William Merle, the weather and time. The weather makes time and the weather makes us, we are made of wind and rain and mark the passing of time with the passing of seasons. I now knew a little more about how it all worked and how we've come to a better understanding of the weather. Up there on the cliff, on a rare occasion when the rain held off me and allowed the weather to bring me sunshine instead, I realised there are few better examples of how the weather is us and the weather is time than the magic in the metronomic daily routine of the solemn, rhythmic intonation of the shipping forecast.

The weather is one of the few things in life that we don't completely understand, can't control and, despite everything, still can't predict with certainty. So much of life is certain these days that it's comforting to think how the major factor in our everyday surroundings, the aspect of our existence that has most influenced our development as a species, is out of our control and resistant to ideology, corruption and politics. I looked out to sea towards the horizon and realised that the shipping forecast, where all the factors and people from this story come together, is the best illustration we have of our relationship with the weather. It's unsullied, honest, respectful, reliable and it helps to keep us safe. Despite the best efforts of a range of people, from geniuses to wackos, we've not been able to tame the weather. What we have been able to do is calibrate our response to it: the paving stones of the shipping-forecast map and the methodical poetry of its content impose our sense of order on to the weather, a kind of negotiated settlement after centuries of philosophy, observation and scientific research.

I turned my back to the sea and looked inland. Far away on the horizon rain was falling silently over the mountains.